TOTAL QUALITY
SAFETY
MANAGEMENT
and AUDITING

TOTAL QUALITY SAFETY MANAGEMENT and AUDITING

Michael B. Weinstein

LEWIS PUBLISHERS

Boca Raton   New York

Library of Congress Cataloging-in-Publication Data

Weinstein, Michael B.
    Total quality safety management and auditing : your complete guide
to developing and auditing a safety management system / Michael B.
Weinstein.
        p.   cm.
    Includes bibliographical references and index.
    ISBN 1-56670-283-6
    1. Industrial safety– –Management.. 2. Industrial safety– –Auditing.
3. Total quality management.   I. Title.
T55.W45   1997
658.3′.82—dc2   97-21402
                                                                                          CIP

No claim to original U.S. Government works
International Standard Book Number 0-8493-283-6
Library of Congress Card Number 97-21402
Printed in the United States of America 1 2 3 4 5 6 7 8 9 0
Printed on acid-free paper

THE AUTHOR

Michael B. Weinstein is the founder and principal consultant for **TLP** (Total Loss Prevention) **Associates**, a consulting practice specializing in safety, loss prevention and risk management. Mr. Weinstein received a M.S. degree in engineering from Case/Western Reserve University and a B.Ch.E. degree from the City College of New York. He has over 30 years of experience in consulting, loss prevention and engineering.

TLP Associates advises on incorporating TQM and ISO-9000 quality management principles into safety and loss prevention programs. Mr. Weinstein's articles on adapting TQM and ISO-9000 to safety have appeared in *Professional Safety* and in *Occupational Hazards*.

As technical director for the engineering department at American Nuclear Insurers, Mr. Weinstein directed development of loss prevention guidelines/criteria for areas such as environmental risk, radiation protection, fire protection and equipment protection (primarily turbine-generators and main transformers). He also managed development of an advanced inspection program, managed a support engineering group responsible for risk and performance assessment programs and engineering support functions, and managed and coordinated varied technical and professional activities.

As a loss prevention professional, Mr. Weinstein has managed and/or conducted loss prevention assessments at all U.S. commercial nuclear facilities focusing on liability/environmental safety, the quality of operations, maintenance and engineering, and management effectiveness.

Mr. Weinstein's engineering experience includes ten years with NASA, at the Lewis Research Center in Cleveland, Ohio. During this period, he was responsible for testing Apollo/Gemini fuel cell system components and for research on advanced nuclear power fuels.

He has published technical reports and journal articles, and has made presentations at many technical conferences, workshops and professional meetings. He is a member of the American Society of Safety Engineers, and is on the Executive Committee of the Central Connecticut Chapter. He also served on the American Nuclear Society's Containment Leakage Testing Standards Committee.

TABLE OF CONTENTS

Chapter 3

Chapter 4

Chapter 5

Chapter 1

TOTAL QUALITY SAFETY MANAGEMENT

I. INTRODUCTION

This is a reference which enables a safety executive or manager to determine if their organization's safety management system incorporates the most modern business, quality, and safety concepts. Included are information and self-assessment questions on 78 quality and safety management topics in four major areas. Specifically, the questions seek to determine if:

- **Total Quality Management (TQM)** principles,
- **ISO-9000** quality guidelines and the
- OSHA **Voluntary Protection Program (VPP)**
- and **Process Safety Management (PSM)** guidelines

are built into the organization's safety management system.

II. SAFETY MANAGEMENT

This book helps to bridge the gap between the theory of quality management and its practical application to safety. For any hazard or for any industry, a complete safety management system must meld hazard or industry-specific technical requirements with key business, quality, and safety management concepts and principles. This book presents those key quality and safety management principles and then sets forth questions which allow managers to assess how well each of the key principles is used in their own safety management process. In specific, the questions in this book are directed toward a **worker or occupational safety program** although the general ideas presented can be adapted to any safety program.

Four sources are used to arrive at the 78 key quality and safety management concepts and principles presented in this book. First, the 24 most important concepts, techniques, and implementation requirements that make up Total Quality Management are identified. Then, the 20 quality program requirements of ISO-9001, the main standard in the ISO-9000 family of standards, are identified.[1] Next, the 20 safety management principles used in OSHA's Voluntary Protection Program are outlined.[2] Finally, 14 hazard management principles set forth in OSHA's Process Safety Management guidelines are also identified.[3]

These 78 quality and safety management concepts and principles are general and can be adapted and used in any safety management system. Combining these quality and safety management principles with any hazard or industry-specific technical requirements will help arrive at a safety management system which incorporates the best business management and quality characteristics. The fully developed safety system focuses on minimizing the number and severity of accidents, continuously improving and achieving a culture of full management and employee involvement in safety.

Behavior-based safety is a systematic process which has natural ties to quality management and has proven beneficial in improving safety programs. Appendix A ties behavior-based safety to the TQM concepts which have been herein identified, and can be used to explore the application of behavior-based safety to the safety management process.

III. SAFETY PERFORMANCE

U. S. Department of Labor statistics[4,5,6] show that occupational injuries and illnesses continue to take an immense toll among private sector workers in the United States. In 1994, the latest year for which complete and analyzed information is available, a total of 6.8 million injuries and illnesses were reported in private industry, a rate of 8.4 for 100 full-time workers. Of these 6.8 million, about 2.2 million resulted in at least one day away from work beyond the date of the injury or illness. Additionally, in 1994 a total of 6,588 fatal injuries were reported, a rate of more than 27 per workday.

Adding to this toll of work injuries and illnesses are the injuries which occur off-the-job. The National Safety Council[7] estimates that in 1995 workers suffered 1.75 off-the-job disabling injuries for every on-the-job injury. In addition, workers suffered almost 8 times the number of fatalities off-the-job compared to on-the-job, 40,400 vs. 5,300. Although off-the-job injuries do not result in direct costs, the indirect costs to industry of hiring and training replacement workers and of having hurt workers at less than full potential, are immense.

In terms of trends, the rate of 8.4 was lower than the previous two years (8.9 in 1992 and 8.5 in 1993), but the rate remains above the range of 8.3 to 7.6 achieved during the years 1980 through 1987. The lowest rate of 7.6 was reached in 1983, thirteen years ago. Overall, it is abundantly clear that no significant improvement has taken place in the rate of injuries and illnesses since 1980 (see Figure 1). Safety effectiveness based on the traditional-style safety management commonly employed is not improving.

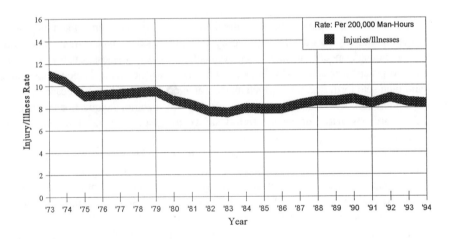

Figure 1. Occupational Injury and Illness Rates in the United States, 1973 - 1994.

Petersen[8] concurs in this conclusion. His analysis of injury statistics from 1973 through 1991 "shows no improvement in lost-time injuries, an 11-percent increase in total recordables, and a 47-percent increase in days lost." Using control charts, Petersen reports that while data from 1981 to 1991 show no significant change (improvement) in terms of total cases, and lost-time cases, there is a significant deterioration in lost work day statistics (evident from 1983 through 1992).

The overall economic and social costs of occupational injuries are immense. The National Safety Council estimates that the true national cost of occupational deaths and injuries was $119.4 billion in 1995. This includes wages and productivity costs, medical costs, administrative costs, and other employer costs. The total lost time away from work due to injuries was 120 million days - equivalent to 60,000 full-time workers for the entire year.[9]

IV. TRADITIONAL SAFETY PROGRAMS

Hansen[10] thoroughly reviews the failings of most safety management programs. He describes safety as most often isolated from the mainstream of an organization and left to "staff managers who lack the authority and organizational position to effect change." As a staff function, safety management is limited in its "ability to identify and resolve management oversights that contribute to accident causation." To succeed, safety must be viewed as a line management function with the line managers responsible and accountable for performance.

The fact is that most companies are compliance oriented. Veltri[11] reports that 77% of companies he surveyed focused on regulatory compliance and traditional safety inspections, and strove for minimal safety investment. In these companies he found that the objective was to avoid problems, not to avoid accidents or promote safety. Only 7% of the companies surveyed desired to elevate their safety management programs to a level of excellence.

Another traditional safety program failing is that the major elements of traditional safety programs do not really work as assumed. In reporting on research into safety and management programs conducted by the National Institute for Occupational Safety and Health (NIOSH), Hansen[12] describes many traditional safety program elements that were found to not correlate with safety effectiveness and results in terms of safety incidents. In some cases, the best safety committees and safety staff were found in the companies that tended to have the higher incident rates. Those traditional safety program elements were: safety committees, safety staff, safety meetings, safety training, safety inspections, safety rules, and safety records. The factors that were found to correlate with better incident statistics were management and cultural factors such as: management involvement, financial support, management - union relations, attitude toward employees, supervisor interactions, planning, and job quality.

Another type of safety program failure, top-down management, is also pointed out by Blair[13] who describes what he terms as the traditional safety paradigm, where management "makes all decisions, establishes the rules and is responsible for making changes when necessary." In this type of environment, what the employees know about the work environment and safety problems is not factored into decision making and problem solving. In his article, Blair states that companies should shift to a paradigm of authentic caring which includes the qualities of active listening, encouraging self-efficacy, removing barriers, and safety coaching.

Hansen[10] cites another basic traditional safety failure of pushing for organizations to change from the traditional wisdom of blaming problems on people, such as "accidents are the result of unsafe employee acts and behaviors," to what he calls the uncommon logic of "accidents are the result of flawed management values, decisions and practices." In his article, Hansen gives examples of traditional safety thinking and how it should be replaced by making basic and continuing change. His recommendations include: finding and fixing basic organizational problems if accident rates are high, correcting poor management attitudes rather than looking for poor worker attitudes, recognizing that management is responsible for 94% of all organizational outcomes (including accidents), not relying on rote obedience to safety rules which can never address all the hazards of a dynamic organization, addressing safety problems by careful up-front planning rather than after-the-fact inspection, and promoting accident analyses that look for the real causes of accidents which are often deeply imbedded in the organization.

Krause and McCorquodale[14] describe another traditional safety failure which is the basic premise of traditional safety incentive programs. In their eyes, the use of safety incentive programs is the result of a basic failure to manage safety, and the results of such programs are very often harmful. Incentive programs tend to: address the wrong behavior - such as not attending safety meetings because they are boring, unjustly reward groups that evidence good results just because of statistical variation, and drive injury and accident reporting underground. The fix is to emphasize the basic value of safety.

V. THE MODERN SAFETY MANAGEMENT SOLUTION

The basic needs for improved safety and safety management can only be addressed through a modern, effective and consistent approach that is applicable to all safety hazards and exposures. The approach offered here is to use the basic concepts, characteristics, techniques, and requirements that make up TQM as one set of building blocks on which to base safety programs. To TQM are added the quality program requirements of the ISO-9000 series of standards and the safety management principles embodied in OSHA's VPP and PSM guidelines to create a comprehensive safety management system (see Figure 2).

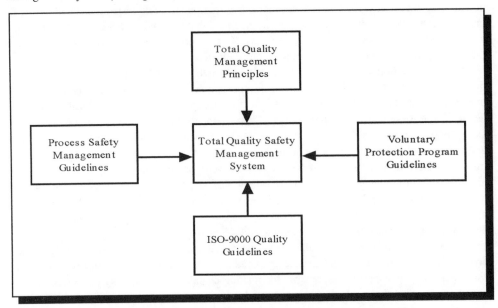

Figure 2. Creating a Total Quality Safety Management System

Capezio and Morehouse[15] have defined TQM as "a management process and a set of disciplines that are coordinated to ensure that the organization consistently meets and exceeds customer requirements. TQM engages all divisions, departments and levels of the organization. Top management organizes all of its strategy and operations around customer needs and develops a culture with high employee participation. TQM companies are focused on the systematic management of data in all processes to eliminate waste and pursue continuous improvement." Over the past years, TQM and ISO-9000 concepts, characteristics, techniques and requirements have gained worldwide acceptance in all types of manufacturing and service industries, and implementation of TQM systems has helped transform many companies by increasing competitiveness, effectiveness and productivity. TQM methods have also been used in smaller doses to foster improvements in specific areas of need such as in customer relations and participation, process control methods, and employee teamwork.

TQM has been proven to be applicable to the practice of safety management as long as the important principles and ideas that foster TQM success are applied for safety success. In a review of why quality processes sometimes fail when used for safety, Manuele[16] reports that efforts to improve safety succeed only when there is: cultural change, real management commitment, achievement of customer satisfaction, the building of safety into systems at the beginning, continuous improvement of processes and systems, the use of statistical process controls, root cause identification of accident causes, responsibility placed on management, employee contribution, and involvement, effective training, and a program of continuous review and feedback. All of these elements are basic to the TQM process and will be discussed in more detail in Chapter 2.

Roughton[17] also describes the applicability of TQM to safety. He focuses on the key elements of establishing a safety culture, using a fully participative management style, developing and really using teams, applying real measurement systems, and providing a firm foundation of effective training. In regard to the supportive safety culture, Roughton states that "in an organization dedicated to safety, management understands that injuries to workers represent an unacceptable waste of resources, and the concept of prevention should flourish." He firms up his call for a positive safety culture with the need for an ethical work environment, consistent and meaningful communications, and a proactive accident prevention mentality.

Safety excellence based on the use of TQM concepts is a fact. As reported in Industrial Safety and Hygiene News,[18] the Coors Brewing Company's Shenandoah brewery set an industry record of 2,701,000 hours without a lost-time accident. This achievement was attributed to a safety process emphasizing individual responsibility, peer safety coaching, and team safety coordinators who promote safety at all levels.

Lark,[19] Creek,[20] and McClay[21] among others, also show the benefits of adopting TQM concepts to safety management. However, although the use of TQM for increased safety effectiveness is laudable, these efforts are not widespread enough, and they certainly have not gone far enough to significantly affect overall industry safety results. Hansen reports that companies and safety professionals have proven slow to accept and apply TQM methods to safety.[22]

Recently, the processes of corporate re-engineering, wherein companies redesign or eliminate corporate resources has begun to negatively affect safety performance. In re-engineered organizations, employees trying to work harder, find their roles are poorly defined and the level of personal security has decreased. These factors lead to an increasing number of accidents. Witherill and Kolak[23] cite five key factors which increase risk during re-engineering processes. These are: fear entering the workplace, employees becoming confused and frustrated, declining morale, employees overwhelmed with increasing work loads, and accountability for safety being lost. Recognition of these factors, up-front, is required to effectively plan for good safety performance in a time of downsizing and re-engineering.

VI. THE BASIC TOTAL QUALITY MANAGEMENT CONCEPTS, TECHNIQUES, AND IMPLEMENTATION REQUIREMENTS

Just what are the basic Total Quality Management concepts and methods? Unfortunately there is no uniformly accepted listing of TQM ideas. Although the number of books and articles describing TQM and its applications seems endless, each author or group has differing ideas on the specifics of what it covers and contains. Nonetheless, Weinstein[24] has described a **practical synthesis** of what TQM means based on the work of established generalists or organizations (for example, Clemmer,[25] Creech,[26] Harrington,[27] Hradesky,[28] and the Malcolm Baldrige Quality Award Guidelines[29]).

The results of this practical synthesis are outlined below.

The basic TQM concepts are:
- Product and Customer Focus
- Leadership Commitment
- Company Culture
- Effective Communication
- Organizational and Employee Knowledge
- Employee Empowerment
- Employee Responsibility and Excellence

- Management by Fact
- Long Range Viewpoint

The important techniques used by TQM organizations are:
- Statistical Process Control
- Structural Problem Solving
- The Best Techniques
- Continuous Improvement
- Quality Management
- Quality Planning

And the steps needed to develop and install a TQM system are:
- Assessment and Planning
- Implementation and Organization
- Cultural Change
- Recognition and Reward Systems
- Leadership Development
- Team Building
- Hiring and Promoting Practices
- Management Readiness
- Total Quality Training

In Chapter 2, these core Total Quality Management principles, techniques and implementation requirements are fully described and adapted to safety management. This entire process from identification to safety management adaptation is outlined in Figure 3.

When TQM is applied to safety, the result is elevating the safety program to a higher level of achievement. This higher level is one of self-motivated, proactive leadership and improvement by all individuals: executives, managers, supervisors, and workers. In such an organization, safety is sought not because of possible fines or penalties, or because the boss or inspector wants it. Nor is safety sought because regulations and standards require it, it is achieved because it is the right thing to do morally, ethically and humanly.

This new higher level of safety achievement is depicted as Level IV in Table I. In this table, the first level of safety programs, Level I, is the level of fear and minimal compliance. The motivation in Level I programs is to avoid fines and penalties and accomplish this by doing as little as possible. Level I safety programs evidence nonexistent or the most minimal safety policies and procedures along with regressive (one-way and fear-based) communication and developmental practices.

Next, Level II is the very traditional, reactive safety program where the motivation for performance is the fear of a poor audit or inspection (or a high insurance premium). The work objective is to follow procedures, as long as they don't overly interfere with the job or production. At this level, upper management is not very interested in safety; the organizations tend to be very centralized with tight control practices; any safety committees are driven by one or just a few powerful members; safety policies are very narrow; recognition and reward systems work poorly; and there is little effort on human development. Level III represents the more modern, professionally-led and active safety program. In a Level III program the workers are taught correct methods and behaviors and they understand what the safety standards mean and what they are for. Level III performance is better-than-average but not necessarily improving. In Level III safety programs, compliance with requirements is the goal and efforts are made to do everything that is required, but no more. Level III programs are characterized by good written safety policies that are not really implemented; well-written safety programs; recognition and reward systems that promote conformity; established downward communication processes; and accident analyses based on compliance adequacy.

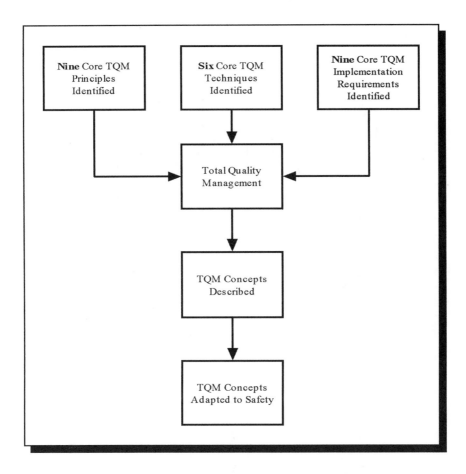

Figure 3. The process of adapting Total Quality Management to safety.

This Level III safety program is similar to the traditional safety programs discussed by Smith[30] and contrasted with a Total Quality approach. He defines traditional safety programs as ones in which safety professionals manage to meet safety specifications. These programs are established to meet company safety rules and government regulations and achieve a modest level of improvement. In these cases, much of the professional's time is spent just providing mandated training and performing routine safety inspections and safety audits. Smith concludes that these programs are intrinsically targeted to just meeting the status-quo, not showing continuous improvement through the correction of real accident causes.

Finally, Level IV is the level of excellence, where the entire organization is passionately committed to safety and betterment. Workers and managers strive together for improvement in an atmosphere of trust and fact. In Level IV programs, there are long-term strategies to develop resources and methods; input to policies comes from all work levels; safety systems consistently exceed regulatory requirements; recognition and rewards are based on acceptance and promotion of the organization's safety values and culture; personal growth is the norm; accident analyses strive for root cause determination and eventual correction; and the entire safety system strives for excellence.

As described by Smith,[31] in a Total Quality safety program, all employees are involved. Employees monitor and chart progress using statistical analysis methods. They function on safety teams which study problems and recommend solutions using quality analysis methods. Management

Safety Level	Motivation	Description	Typical Assessment Method	Typical Learning Method	Typical Safety Goal	Typical Safety Result
I	Fear	Inactive	OSHA Inspection Only	Basic Required Training Only	No Fines, Penalties	Less Than Full Compliance, Worse Than Average Record
II	External Punishment	Reactive	Paperwork Audit, Inspection	Classroom Instruction, Testing	No Noncompliances, Citations	Rote Compliance, No Improvement, Average Record
III	External Reward	Active Understanding and Belief	Work Observation	In-depth Instruction, Coaching	All Jobs Done Correctly	Appropriate Behaviors, Better Than Average Record
IV	Self and Internal	Proactive Passion and Commitment	Peer and Subordinate Interviews, Work Results	By Example, Self-Learning	No Accidents, Best Methods	Continuous Improvement and Leadership, Excellent Record

Table I. Levels of Safety Management Possible in an Organization

focuses on solving problems by changing the management controlled safety systems. The eventual goal is for 100% accident reduction.

VII. THE ISO-9000 STANDARDS

The ISO-9000 family of standards represents the common denominator of business quality that is accepted internationally. These standards were first adopted in 1987. Certification demonstrates the capability of a supplier to control the processes that determine the acceptability of a product or service being supplied. ISO-9000 is product quality oriented, it does not cover the efficiency of the organization or other factors that may be part of a complete Total Quality Management program.

There are a number of separate guidance and conformance standards in the ISO-9000 family. The published and draft standards are:

- **ISO-9000-1**Quality management and quality assurance standards - Part 1: Guidelines for selection and use (a guidance standard). **ISO-9000-2: 1993** Quality management and quality assurance standards - Part 2: Generic guidelines for the application of ISO 9001, ISO 9002 and ISO 9003. **ISO-9000-3: 1991** Quality management and quality assurance standards - Part 3: Guidelines for the application of ISO 9001 to the development, supply and maintenance of software. **ISO-9000-4: 1993** Quality management and quality assurance standards - Part 4: Guide to dependability program management.

- **ISO-9001: 1994** Quality Systems - model for QA in design/development, production, installation and servicing (the most complete and rigorous conformance standard). **ISO-9002: 1994** Quality Systems - a model for QA in production, installation and servicing (a conformance standard). **ISO-9003: 1994** Quality Systems - a model for QA in final inspection and test (a conformance standard)

- **ISO-9004-1: 1994** Quality management and quality system elements - Guidelines (a guidance standard). **ISO-9004-2: 1991** Quality management and quality systems elements - Part 2: Guidelines for services. **ISO-9004-3: 1993** Quality management and quality systems elements - Part 3: Guidelines for processed materials. **ISO-9004-4: 1993** Quality management and quality systems elements - Part 4: Guidelines for quality improvement

- **ISO-10005: 1995** Quality Management Guidelines - Guidelines for quality plans. **ISO/DIS-10006:** Quality Management Guidelines - Guidelines to quality in project management. **ISO-10007:** Quality Management Guidelines - Guidelines for configuration management.

- **ISO-10011:** Guidelines for auditing quality systems. **ISO-10011-1: 1990** Guidelines for auditing quality systems Part 1: Auditing. **ISO-10011-2: 1991** Guidelines for auditing quality systems Part 2: Qualification criteria for quality systems auditors. **ISO-10011-3: 1991** Guidelines for auditing quality systems Part 3: Management of audit programs.

- **ISO-10012-1: 1992** Quality Assurance requirements for measuring equipment - Part 1: Metrological confirmation system for measuring equipment. **ISO/FDIS-10012-2:** Quality Assurance requirements for measuring equipment - Part 2: Control of measurement processes.

- **ISO-10013: 1995** Guidelines for developing quality manuals.

- **ISO/DIS-10014:** Guidelines for managing the economics of quality.

When considering the ISO-9000 quality requirements, the ISO-9001 standard is used as the model for a quality system since it is the most complete rigorous conformance standard. ISO-9002 and ISO-9003 contain fewer requirements and are essentially subsets of ISO-9001.

The ISO-9000 standards focus on 20 aspects of a quality program that is subject to rigorous auditing during certification. These aspects (from ISO-9001) are listed below. In this book these 20 aspects of quality are considered as general management concepts and are adapted and applied to safety management.

The 20 ISO-9000 quality program requirements are:
- Management Responsibility
- Quality System
- Contract Review
- Design Control
- Document and Data Control
- Purchasing
- Customer Supplied Product
- Product Identification and Traceability
- Process Control
- Inspection and Testing
- Control of Inspection, Measurement and Test Equipment
- Inspection and Test Status
- Control of Nonconforming Product
- Corrective and Preventive Action
- Handling, Storage, Packaging, Preservation and Delivery
- Control of Quality Records
- Internal Quality Audits
- Quality Training
- Servicing
- Statistical Techniques

VIII. THE OSHA VOLUNTARY PROTECTION PROGRAM

The OSHA Safety and Health Program Management or Voluntary Protection Program Guidelines provide guidance for companies striving for excellence in safety. They have been implemented by OSHA to encourage employers to reduce hazards, institute new programs, and perfect existing health and safety programs. When approved, participants in the VPP are exempt from the normal random and programmed OSHA inspections but instead company performance records are submitted for review and the company is subject to a periodic reverification every three years. OSHA approval for a VPP site shows not only adherence to the letter of the standards but a demonstrated management commitment to address all unsafe and hazardous conditions.

There are three levels of company participation in the VPP which are awarded once the company has submitted an application and successfully completed a detailed on-site evaluation. The on-site evaluation verifies the accuracy of material in the application, identifies strengths and weaknesses in the safety program, and determines if the safety program is effective in protecting against site hazards.

- Star status is given to companies which have instituted comprehensive safety and health programs and have been successful in reducing hazards and demonstrating good injury rates.

This award recognizes that the company is a leader in developing innovative methods to prevent injury and illness. Companies earning the Star status can use the VPP Star logo to promote themselves.

- Merit status is given to companies which have the potential and commitment to attain Star status. These companies have demonstrated high achievement in safety and health, but must still make improvements to graduate to Star status. Merit facilities are evaluated annually

- Demonstration status shows that companies have attained Star or Merit quality using alternative approaches to safety. This status is generally awarded to small businesses or construction companies. Demonstration facilities are also reevaluated annually.

As of April 1995, 82 companies with 189 separate sites were involved in the VPP. Of these, 148 separate sites had attained Star status, 40 sites had received Merit status and one site had Demonstration status. These companies employ approximately 153,000 workers.[32]

VPP results are excellent. OSHA reports[33] that 9 of the 178 sites in the program as of 1994 had no incidence of injuries. Overall, the sites showed only 45% of the injuries expected based on comparable industry experience. Similarly, 31 of the 178 sites had no lost workday injuries. Overall, the number of lost workday injuries was 51% below the average expected for similar industries. Although these results are superb, they have not led to a corresponding decrease in national injury and illness rates due to the limited acceptance of the VPP.

The VPP has also resulted in decreased workers' compensation costs and examples of increased production and product quality and improved employee morale. Workers' compensation costs at several sites have declined 70 to 89% since commencing the VPP process. Other benefits cited have been improved communication between employees and management, and positive public relations.

Although these results are superb, the VPP has not had an impact on national injury and illness rates because of its very limited acceptance. Although 159,000 workers are covered, this only represents approximately 0.2% of the national workforce, a number too small to impact the national injury rate of 8.4 per 100 full-time workers.

Partly in an effort to foster the widespread implementation of effective safety and health programs, and to have a greater impact on national injury rates, OSHA is preparing to embark on a series of individual state initiatives, known as the Cooperative Compliance Programs.[34] Based on site-specific injury and illness data requested in 1995, OSHA will direct its resources to those sites with the highest rates of serious injuries and illnesses. In this cooperative program, the sites will be asked to focus on the development of comprehensive, effective safety and health programs, which follow the general Voluntary Protection Program guidelines.

In 1996, OSHA adopted revised VPP guidelines to replace those originally enacted in 1989.[35] Among the changes were significant additions regarding contract workers and the need to evaluate Process Safety Management programs for sites where they are required.

An outline of the topics covered in the VPP guidelines is given below. As are true for the TQM and ISO principles and requirements presented earlier, these VPP topics are general and are applicable to any safety management system. In this book, the emphasis is on worker and occupational safety.

The topics in the Voluntary Protection Program guidelines are:
- Safety Management Program
- Safety Policy and Objectives
- Management Commitment and Involvement
- Employee Participation

- Assigned Responsibilities and Accountabilities
- Contract Workers
- Program Review
- Work-site Hazards Analysis
- Baseline Surveys
- Site Inspections
- Hazard Prevention and Control
- Procedures and Protective Equipment
- Facility and Equipment Maintenance
- Hazard Communications
- Accident and Injury Analysis
- Emergency Planning
- Medical Program
- Safety and Health Training
- Safety and Health Training Content
- Supervisor and Manager Readiness

IX. PROCESS SAFETY MANAGEMENT

The Process Safety Management guidelines are performance-based management controls intended to provide a systematic approach to managing and evaluating the whole of a chemical process. As promulgated by OSHA, the PSM guidelines are specifically directed to processes involving more than 130 listed (highly hazardous) chemicals in specified quantities and flammable liquids and gases in quantities exceeding 10,000 pounds.[36] Citing disasters in Bhopal, India and Pasadena, Texas and others, OSHA stated that hazardous chemical releases pose a significant threat to employees (and the public). The proposed guidelines were issued for comment on July, 17, 1990 and after numerous comments and hearings, published as final on February 24, 1992. It is estimated that from 25,000[37] to 88,000[38] facilities are covered by the PSM Rule. An outline of the topics covered by the Process Safety Management guidelines is shown below.

The topics covered in the Process Safety Management guidelines are:
- General Guidelines
- Employee Involvement
- Process Safety Information
- Process Hazards Analysis
- Operating Procedures and Practices.
- Employee Training
- Contractors
- Pre-Startup Safety
- Mechanical Integrity
- Nonroutine Work Authorizations
- Managing Change
- Investigation of Incidents
- Emergency Preparedness
- Compliance Audits

As published, the OSHA standard applies to processes involving highly hazardous chemicals and mainly to manufacturing industries - particularly those pertaining to chemicals, transportation equipment and fabricated metal products. By a process, OSHA means any activity including using, storing, manufacturing, handling, or moving such chemicals. However, from the standpoint of general safety management, the ideas in the PSM standard can be considered as **an integral part of any safety management system**. Instead of only being directed to chemical processes, the whole range of workplace processes and activities (electrical, mechanical, fluid/pressure, etc.) can be managed and controlled using the same concepts.

Basically, the same requirements for process safety management are followed in the Environmental Protection Agency's (EPA) Risk Management Program.[39] The Clean Air Act requires the EPA to regulate against accidental releases of regulated substances and to reduce the severity of those releases that do occur. After careful consideration, the EPA issued regulations saying that "processes in industry categories with a history of accidental releases and processes already complying with OSHA's PSM Standard will be subject to a prevention program that is **identical** to parallel elements of the OSHA Standard." Thus one accident prevention program will be used to protect workers, the general public and the environment. Under the EPA's rule, processes with lesser potential for impact on the public in the event of an accidental release will either be subject to minimal requirements or to midrange requirements for hazard assessment, prevention, and emergency response. The regulations apply to stationary sources with processes that contain more than threshold quantities of regulated materials. The risk management plans developed under the EPA rule must be registered with the EPA, submitted to state and local authorities, and made available to the public.

In determining what actions and requirements are necessary, owners must conduct a worst-case release analysis considering the release of the largest quantity of a regulated substance from a vessel or process line failure, including administrative controls and passive mitigation that limit the total quantity involved in the release rate. The hazard assessment also considers alternative release scenarios which are accidents most likely to occur.

X. OTHER FRAMEWORKS FOR SAFETY MANAGEMENT

Other outlines for safety management exist and can be used to ensure the completeness and adequacy of a specific safety management system. One such outline is the Process Safety Management Program developed by the AIChE Center for Chemical Process Safety.[40] These guidelines cover 13 general process areas as follows:

- Availability Objectives and Goals: continuity of operations, continuity of systems, continuity of organizations, company expectations, quality processes, control of exceptions, alternative methods, management accountability.
- Process Knowledge and Documentation: process definition and design criteria, process and equipment design, company management information, documentation of risk management decisions, protective systems, normal and upset conditions, chemical and occupational health hazards.
- Capital Project Review and Design Procedures: appropriation request procedures, risk assessment for investment purposes, hazard review, siting relative to risk management, pilot plant, process design and review procedures, project management procedures.
- Process Risk Management: hazard identification, risk assessment of existing operations, reduction of risk, in-plant emergency response and mitigation, process management during emergencies, encouraging client and supplier companies to adopt similar risk management practices, selection of business with acceptable risks.
- Management of Change: change of technology, change of facility, organizational changes that may impact process safety, variance procedures, temporary changes, permanent changes.
- Process and Equipment Integrity: reliability engineering; materials of construction; fabrication and inspection procedures; installation procedures; preventive maintenance;

process, hardware, and systems inspection and testing; maintenance procedures; alarm and instrument management; demolition procedures.
- Human Factors: human error assessment, operator/process and equipment interfaces, administrative controls vs. hardware.
- Training and Performance: definition of skills and knowledge, training programs, design of operating and maintenance procedures, initial qualification assessment, instructor program, records management.
- Incident Investigation: major incidents, near-miss reporting, follow-up and resolution, communication, incident reporting, third-party participation if necessary.
- Standards, Codes and Laws: internal and external standards, guidelines and practices.
- Audits and Corrective Action: process safety audits and compliance reviews, resolutions and close-out procedures.
- Enhancement of Process Safety Knowledge: internal and external research, improved predictive systems, process safety reference library.

Another framework for safety management is provided by the new ISO-14000 standards for environmental management. Although directed to environmental protection, the general guidance and outline of topics covered is adaptable to occupational safety management. The topics covered in ISO-14001[41] are:
- Environmental Management System
- Environmental Policy
- Planning
  - Environmental Aspects
  - Legal and Other Requirements
  - Objectives and Targets
  - Environmental Management Programs
- Implementation and Operation
  - Structure and Responsibility
  - Training, Awareness and Competence
  - Communication
  - Environmental Management System Documentation
  - Document Control
  - Operational Control
  - Emergency Preparedness and Response
- Checking and Corrective Action
  - Monitoring and Measurement
  - Non-Conformance and Corrective and Preventive Action
  - Records
  - Environmental Management System Audit
- Management Review

XI. NOTES ON USING THIS BOOK

This book is formatted such that background and reference material is provided immediately preceding the audit/assessment questions for each of the 78 topics addressed. For consistency, all assessment questions are phrased so that positive responses (yes, always, etc.) are desired. In addressing the questions, it is intended that the responder considers if **all** of the requirements or attributes are met **all** of the time. The answers might be that:
- **all,**
- **most,**

- **about half,**
- **some** or
- **none**

of the attributes or requirements are met. This specificity will help in analyzing the results. In all cases the answers should try to address the intent of the questions, the approaches used, the extent of deployment and application and the results achieved in the area. The three evaluation dimensions of approach, deployment and results are those used in the Malcolm Baldrige National Quality Award Criteria.[42]

- Approach refers to the methods used to address an item: the appropriateness of the methods; the effectiveness of the methods; and evidences of innovation which include the adaptations of approaches used by others.
- Deployment refers to the extent of application of the approach to all requirements of the item and to all appropriate work units.
- Results refers to the outcomes achieved for the purposes given in the item: the performance levels; comparative performance levels; the rate, breadth and importance of improvements; and the demonstration of sustained improvement or high-level performance

To aid in the self-assessment process, forms are provided for summary analysis and reporting at the end of each chapter.

Although the intent of the questions is to enable a self-assessment, the questions can be used as part of a full compliance or effectiveness audit. In that case, the audit should also include observations, interviews, documents and records reviews, and an assessment of implementation results. A more complete discussion of the auditing process is given in Appendix B. This appendix discusses the auditing process first from the viewpoint of ISO standards, then adds material on preparing logically constructed audit plans. Appendix B also has material on interviewing, developing findings, conclusions and recommendations; and on conducting exit interviews.

Since four independent sources (TQM, ISO-9000, VPP and PSM) were used to arrive at the total of 78 key quality and safety management principles presented in this book, there is some overlap of material among the sections. In considering or addressing a major management topic such as employee involvement, accident analysis, or training, the relevant material in each of the sections could be combined. For example, there is material on the topic of accident analysis or investigation contained in the following sections.

- Chapter 2, No. XI: Structural Problem Solving
- Chapter 3, No. XIV: Corrective and Preventive Action
- Chapter 4, No. XV: Accident and Injury Analysis
- Chapter 5, No. XII: Incident Investigation.

Other areas where significant material is found in different chapters of the book are listed below.

Topic	Where Found
Training	Chapter 2, V: Organization and Employee Knowledge
	Chapter 2, XXI: Leadership Development
	Chapter 2, XXIII: Management Readiness
	Chapter 2, XXIV: Total Quality Training
	Chapter 3, XVIII: Quality Training
	Chapter 4, XVIII: Safety and Health Training
	Chapter 4, XIX: Safety and Health Training Content
	Chapter 4, XX: Supervisor and Manager Readiness
	Chapter 5, VI, Employee Training

16

XI. REFERENCES

1. Quality Systems - model for QA in design/development, production, installation and servicing, ISO-9001: 1994, International Organization for Standardization, Geneva, Switzerland, 1994.

2. Safety and Health Management Guidelines, U. S. Occupational Safety and Health Administration, Federal Register, 59:3904-3916, 1989.

3. Process Safety Management, OSHA-3132, U. S. Occupational Safety and Health Administration, Washington, D. C., 1994 (Reprinted).

4. National Census of Fatal Occupational Injuries, 1994, Technical Information Notice USDL-95-288, U. S. Department of Labor, Washington, D. C., 1995.

5. Workplace Injuries and Illnesses in 1994, Technical Information Notice USDL-95-508, U. S. Department of Labor, Washington, D. C., 1995.

6. Characteristics of Injuries and Illnesses Resulting in Absences From Work, 1994, Technical Information Notice USDL-96-163, U. S. Department of Labor, Washington, D. C., 1996.

7. Accident Facts: 1996 Edition, National Safety Council, Itasca, IL, 1996, 54.

8. **Petersen, D.**, The occupational safety and health act of 1970: 25 years later, *Professional Safety*, 41.12, 27, 1996.

9. Accident Facts: 1996 Edition, 51.

10. **Hansen, L.**, Safety management: A call for (r)evolution, *Professional Safety*, 38.3, 16, 1993.

11. **Veltri, A.**, Transforming safety strategy and structure, *Occupational Hazards*, 53.9, 149, 1991.

12. **Hansen, L.**, "Re-braining" corporate health and safety, *Professional Safety*, 40.10, 24, 1995.

13. **Blair, E. H.**, Achieving a total safety paradigm through authentic caring and quality, *Professional Safety*, 41.5, 24, 1996.

14. **Krause, T. R. and McCorquodale, R. J.**, Transitioning away from safety incentive programs, *Professional Safety*, 41.3, 32, 1996.

15. **Capezio, P. and Morehouse, D.**, *Taking The Mystery Out Of TQM*, 2nd Edition, Career Press, Franklin Lakes, NJ, 1995, 1.

16. **Manuele, F. A.**, Quality and safety: a reality check, *Professional Safety*, 40.6, 26, 1995.

17. **Roughton, J.**, Integrating a total quality management system in safety and health programs, *Professional Safety*, 38.6, 32, 1993.

18. News Roundup, *Industrial Safety and Hygiene News*, December, 1995, Internet: www.safetyonline.net/ishn/9512/news.htm.

19. **Lark, J.**, Leadership in safety, *Professional Safety*, 36.3, 33, 1991.
20. **Creek, R. N.**, Organizational behavior and safety management, *Professional Safety*, 40.10, 36, 1995.
21. **McClay, C. J.**, Achieving breakthrough in safety via employee empowerment, *Professional Safety*, 40.12, 44, 1995.
22. **Hansen**, Safety management, 17.
23. **Witherill, J. W, and Kolak, J.**, Is corporate re-engineering hurting your employees, *Professional Safety*, 41.5, 28, 1996.
24. **Weinstein, M. B.**, Total quality approach to safety management, *Professional Safety*, 41.7, 18, 1966.
25. **Clemmer, J.**, *Firing on all Cylinders - The Service/Quality System for High-Powered Corporate Performance*, Business One Irwin, Homewood, IL, 1992.
26. **Creech, B.**, *The Five Pillars of TQM*, Truman Talley Books/Dutton, New York, NY, 1994.
27. **Harrington, H. J.**, *Total Improvement Management*, McGraw-Hill, New York, NY, 1995.
28. **Hradesky, J. L.**, *Total Quality Management Handbook*, McGraw-Hill, New York, NY, 1995.
29. *Malcolm Baldrige National Quality Award 1996 Award Criteria*, American Society for Quality Control, Milwaukee, WI, 1995.
30. **Smith, T. R.**, Will safety be ready for workplace 2000, *Professional Safety*, 41.2, 37, 1996.
31. **Smith**, Will safety be ready, 38.
32. **Baughn, K. T., Libby, B. E., and Rogers, R.**, *Chemical Health and Safety*, 2.5, 8, 1995.
33. Benefits of the Voluntary Protection Program, U. S. Occupational Safety and Health Administration, Washington, D. C., 1994.
34. Connecticut Cooperative Program, U. S. Occupational Safety and Health Administration, Draft Instruction CPL., Draft, Hartford/Bridgeport Area Office, 1996.
35. Revised Voluntary Protection Programs (VPP) Policies and Procedures Manual, Instruction TED 8.1a, U. S. Occupational Safety and Health Administration, Washington, D. C., 1996.
36. Process Safety Management of Highly Hazardous Chemicals, Fact Sheet 92-45, U. S. Occupational Safety and Health Administration, Washington, D. C., 1992.
37. **Dennison, M. A.**, *OSHA and EPA Process Safety Management Requirements*, Van Nostrand Reinhold, New York, NY, 1994, 9.
38. **Stricoff, R. S.**, Safety Analysis and Process Safety Management, in *Risk Assessment and Management Handbook*, Kolluru, R. V., Ed., McGraw Hill, New York, NY, 1966, Ch. 8.
39. Risk Management Program Rule, 40 CFR 68, U. S. Environmental Protection Agency, Federal Register, June 20, 1996, 31668-31730.
40. **Dennison**, *OSHA and EPA Process Safety Management Requirements*, 55-57.
41. Environmental management systems: specifications with guidance for use, ISO-14001:1996, International Organization for Standardization, Geneva, Switzerland, 1996.
42. Malcolm Baldrige National Quality Award 1996 Award Criteria, 24.

Chapter 2

TOTAL QUALITY MANAGEMENT - ADAPTED
TO OCCUPATIONAL SAFETY

In this chapter the concepts, techniques and implementation requirements that make up Total Quality Management are described and adapted to occupational or work safety.

The TQM concepts discussed are:
- Product and Customer Focus
- Leadership Commitment
- Company Culture
- Effective Communication
- Organizational and Employee Knowledge
- Employee Empowerment
- Employee Responsibility and Excellence
- Management by Fact
- Long-range Viewpoint

TQM techniques described are:
- Statistical Process Control
- Structural Problem Solving
- Best Techniques
- Continuous Improvement
- Quality Management
- Quality Planning

The requirements for development and implementation of a TQM management system are:
- Assessment and Planning
- Implementation and Organization
- Cultural Change
- Recognition and Reward Systems
- Leadership Development
- Team Building
- Hiring and Promoting
- Management Readiness
- Total Quality Training

In the material that follows, each of the 24 TQM topics is described. Then each concept, technique or requirement is adapted to a TQM-style safety management system in which the forces of customer focus, leadership commitment, and employee empowerment drive a continuously improving safety system. At the end of each topic, assessment questions are presented which allow evaluation of the extent to which TQM ideas have been adopted in an organization. Answering the questions will permit managers and executives to discover and evaluate weaknesses in their management programs and ideas - and will point the way to potential improvement.

I. PRODUCT AND CUSTOMER FOCUS

Product and customer focus simply means *keeping your eyes on the prize*. For product and service oriented companies, the prize is top quality products or services and fully satisfied customers. For safety, the prize is having no accidents and employees fully satisfied with the safety management system and practices.

A. CUSTOMER FOCUS

Customer focus or customer-driven quality is one of the core set of values and concepts that the Malcolm Baldrige Award Criteria[1] is based on. This concept means that in an organization there must be constant sensitivity to the customer along with measurement of the factors that drive customer satisfaction and retention. Ultimate quality is judged by the customers, and all the product and service characteristics that contribute to customer value must be key foci of the management system. The management system, the managers, and the employees must be customer-directed and attentive, and processes must be in place to learn what the customers want and what they think.

In terms of worker safety, every employee and manager should be viewed as a customer, an internal customer. Safety in terms of no injuries, accidents or work-related illnesses then is the product, output and result of quality processes and activities. Thus the focus of all activities, planning and culture should be on the customer (the worker) and the product (safety). In a customer-driven organization, health and safety services are provided to internal customers.

There are also other customers for safety as described by Smith[2]. These other customers are:

- the employee's family,
- the regulators,
- the subcontractors, and
- other company customers.

The elements of safety as a product are: safety-related processes, procedures, and equipment fit-for-use and defect-free, and a safe work environment as perceived by the customer (worker).

Focusing on the employee as a customer should extend to establishing off-the-job safety programs. Off-the-job injuries are more frequent than at work, and they significantly affect job costs and performance factors such as absenteeism, medical treatment and insurance.[3] Research has shown that off-the-job safety programs are part of effective accident cost control programs.[4]

In focusing on customers, the organization needs to determine and understand who the customers are, what their expectations are, how satisfied they are, and what might be done to better serve them and increase their satisfaction. The customers (workers) are the ones who really define safety for the organization so that safety expectations, perceptions, and suggestions for improvement should be sought from them. Elements of safety to be addressed should include:

- timeliness of response;
- condition of products, equipment, or services;
- availability of services; and
- the keeping of agreements.

B. PRODUCT FOCUS

Product focus is addressed by Creech[5] who states that the product (defined in terms of customer requirements) should be the focal point for organizational purpose and achievement. In an organization, the result of a product mindset is having a "we" orientation compared to having an "I" oriented job mindset. The focus of all employee activities should be on the product and not on the job. Ingredients of good products are: pride, professionalism, purpose, and progress. Quality in the products flows from performance which flows from pride and professionalism.

The safety implications of a product or results orientation is that the organization, as a whole, should strive to meet the established and agreed-upon safety objectives. For example, a viable focus in safety-conscious companies is a real expectation of zero-injuries. Nelson[6] reports that many construction companies have achieved "zero-lost workday case injuries" on individual projects or for a full year. In fact, one project logged 1.2 million hours without a recordable injury. A real commitment to a zero-injury goal by executives, coupled with resources and training, effectively fosters a culture of safety excellence among the workforce.

PRODUCT AND CUSTOMER FOCUS: ASSESSMENT QUESTIONS

- Is a major focus of all company activities, planning, and culture:
  - on the employee?
  - on providing a safe, injury-free work environment and essential services?
  - on meeting employee expectations?
- Do all the executives consider the employees to be customers?
- Have employee expectations and concerns regarding safety been determined in any formal and consistent manner?
- Do the executives know what the employees think about the safety programs?
- Do management and the workers have an expectation of zero injuries and continuous safety improvement?
- Does the safety policy refer to employee and management expectations?
- Does the organization communicate on safety with external safety customers, including the employee's family?

II. LEADERSHIP COMMITMENT

A. QUALITY COMMITMENT

The Malcolm Baldrige National Quality Award Criteria state that leadership means that the company's leaders set directions and the customer orientation, evidence clear and visible values and have high expectations.[7] Leadership commitment is a passionate, focused, sustained effort to define and achieve TQM by senior leaders.

For success, top management must be totally committed and supportive of TQM. The single most frequent cause of TQM failure is the lack of commitment from the CEO and other senior staff. Leadership commitment can be measured by the leaders' TQM accomplishments, by their involvement in quality improvement actions, and by the percent of time they spend on TQM issues. Leadership commitment is said to transform promise to reality. Similarly, an organization will not produce safety excellence without full leadership commitment to safety.

Clemmer[8] cites signaling commitment as the most important step in the process of culturally changing to a TQM organization. Senior management signals by its daily behavior that service/quality improvement is a true cultural shift, not just another program. Senior executives lead by visible example and they are team leaders. They lead in the change process by learning new team-based skills and service/quality improvement techniques. Managers show commitment by getting out of their offices, and by getting around and involved with the work.

Clemmer defines five levels of increasing executive and management commitment.[9] These levels are:
- permission - which means just supporting if it doesn't cost too much;
- lip service - or giving speeches and writing memos with little budget or support;
- passionate lip service - where the key executives briefly attend some training and some

elements of deployment are in place;
- involved leadership - where leaders are involved and leading for safety and all elements of the safety deployment process are in place and working;
- strategic service/quality leadership - in which day to day operating decisions are delegated to team levels, the majority of an executive's time is involved with customers, suppliers, teams, and managers in gathering information.

Twelve key processes are discussed by Clemmer as measures of the management teams' commitment to quality.[10] In terms of indicators these are:
- Personal time commitment - investing a substantial amount of personal and organizational time over an extended period to quality improvement.
- Managements' accountability - holding line managers accountable for quality improvement.
- Progress review - regularly reviewing and reinforcing quality improvement efforts.
- Resource provision - committing the financial and human resources needed for full deployment.
- Integration - integrating improvement efforts with strategic and financial planning.
- Personal and organizational change - revising personal habits or organizational systems and processes that hinder quality improvement. This includes better time management, and improved delegation, decision-making, and meeting skills.
- Team use - bringing employee teams heavily into the improvement planning and implementation process.
- Personal education - investing 10 to 12 days per year in personal education, learning, and skills development.
- Feedback - seeking continuous feedback on how well management is signaling its quality vision and core values.
- Training support - ensuring that coordinators and trainers have plenty of training and highly visible support.
- Personal involvement - personally leading committees and teams and using data-based tools and techniques in decision making.
- Process maintenance - maintaining a steady and continuous stream of education and awareness.

Taken together, these processes demonstrate that the management team is providing their full interest and support to the quality effort. They are showing by their actions that quality is important to them, and they are insisting on it being important to everyone.

Harrington[11] believes that the commitment by a top executive of personal time has the greatest impact on the improvement effort. This use of a vital resource, time, signals the leader's full dedication to quality. Another indicator of leadership commitment is tying one's compensation to improvement measurements.

B. SAFETY COMMITMENT

Two important indicators of commitment to safety are the quality and extent of plant and equipment maintenance and the weight that production pressure has on safety considerations. Krouse[12] reports that workers see poor maintenance as evidence of a low commitment to safety. If production is not stopped because of a safety problem, management's lack of commitment to safety is made clear to everyone.

Nelson[13] discusses commitment to safety in terms of achieving a zero-injury culture in construction companies. His recommendations for organizations hoping to get to this performance level include:
- Top management must evidence a passion for zero injuries.
- Management ensures that line managers are responsible for safety performance.

- Senior management meetings are begun with emphasis on safety performance. The CEO gets regular safety reports.
- Top management (the CEO) gets immediate notification of lost-workday injuries.
- Senior managers immediately investigate lost-workday accidents and help plan for corrective/preventive action.
- Senior management is involved in recognition for zero-injury achievement.
- Managers are evaluated on safety performance.

Of course, management must also ensure that sufficient safety resources are available. In terms of the 7 M's described by Harrington,[14] this means:
- Men: All employees are properly trained and proficient in safety and work skills.
- Machines: All equipment is reliable, well maintained and properly safeguarded.
- Methods: All work and safety processes use the best appropriate technologies and all procedures are verified safe and effective.
- Materials: All components, ingredients, and documentation are correct, and properly described and labeled.
- Media: There is sufficient time afforded to safety processes and to safety learning, and the environment is supportive.
- Motivation: All employees want to succeed and improve, and this attitude is reinforced by recognition and reward systems.
- Money: Full financial support is given to safety efforts. Safety training and personal growth are supported.

LEADERSHIP COMMITMENT: ASSESSMENT QUESTIONS

- Are all our executives actively and fully committed and working toward a superior work safety culture?
- Do all our executives demonstrate their commitment and involvement on a routine basis?
- Do all executives really know about the company's safety programs, safety record, and safety activities?
- Are managers clearly responsible and accountable for safety performance?
- Do managers follow all the safety rules and requirements all the time?
- Do senior managers investigate accidents and conduct safety inspections, reviews, and audits?
- Do the executives provide the safety function with the resources needed (men, machines, methods, materials, media, motivation, and money) to promote and improve safety?
- Do all the managers:
  - go through the safety training?
  - with their groups?
- Does the CEO or other top management get immediate notification of accidents?
- Do top management meetings include a review of safety performance?

III. COMPANY CULTURE

In a quality culture, values must be in place to produce the behaviors that prevent quality problems. In any organization, the culture is determined by the business environment, the values, the heros, the rites and rituals, and the cultural network. Cultures can be strong (where there is open discussion of the values and no tolerance of deviance) or they can be weak. For a TQM organization to succeed, the TQM culture must be strong. Likewise for safety to succeed, the safety culture must be strong, and safety must be a company value.

TQM cultures are also exemplified by the presence of sincerity and trust. These elements turn into direct savings in terms of fewer written notes and contacts, and fewer inspections and audits performed just to ensure compliance and solidify agreements. Trust fosters a more oral culture rather than a written one. The tasks of documenting and authorizing activities waste energy that should be involved in actually doing the work.[15]

Trust is also important in fostering organizational change.[16] For example, without employee trust in management, any attempts to implement new procedures and processes will be hindered by the level of unacceptance on the employees' part. The employees will not accept that the procedures need to be changed and that the changes proposed are correct. If trust is lost, it can be rebuilt through a process of enhanced communication and repeated positive reinforcement.

A company must possess the vision of a safe work environment. (Hradesky defines vision as the state of "being" that the company desires to achieve in 3-5 years.)[17] Company values and behaviors (management and worker) must be established to promote worker safety. There should be a strong safety culture established with no tolerance for unsafe practices.

This point is further exemplified by Kelly,[18] who states that safety should be treated as a human value, not a company priority. Where safety is a value, management never condones unsafe acts or conditions. They communicate their safety will by actions - always making their commitment to safety clear.

With this vision, all values - the beliefs, attitudes and behaviors of the organization that are observable and are required to achieve the vision must be consistent. (Values drive attitudes (what we feel and believe) which drive behaviors (what we say and do).) In an organization dedicated to safety, management understands that injuries to workers represent an unacceptable waste of resources.[19]

Assessment of the safety culture includes such ideas as:
- Is there a strong safety culture established with no tolerance for unsafe practices?
- Is the cultural goal zero injuries?
- Are health and safety procedures followed, all the time?
- Is there a vision of a safe work environment and do all employees share in it?
- Do employees value safe behavior, themselves, and their continued well-being?
- Are the management style and culture non-autocratic with a win-win atmosphere?
- Is there a trusting relationship between management and employees?
- Do employees believe that safety is a company priority?

In organizations with a strong safety culture:
- executives and managers visibly support safety with no contradictory decisions and full accountability,
- employees are involved with safety and their views are sought and acted on, and
- supervisors' actions support safety including recognizing and appreciating safe work practices and behaviors.[20]

An important element in a positive safety culture is ethical integrity.[21] Ethical integrity means that promises are kept, contracts (implied and formal) are adhered to, and the organization and managers see that safety is vital. In organizations with ethical integrity, managers perceive and are sensitive to safety issues, they comply with safety standards, they involve the entire work community in safety processes, and they resolve safety problems in a principled and consistent way. These managers are role models who enhance the credibility of the entire safety management system.

Other functions of the ethical manager include:
- ensuring the organizations' character and reputation are aligned through accurate communications;
- responsibly working to initiate and accept safety changes; and
- valuing others for their inherent worth.

COMPANY CULTURE: ASSESSMENT QUESTIONS

- Are company values, attitudes, and behaviors (management and employee) established to promote employee safety?
- Is safety a value at this company?
- Is there a strong safety culture established with no tolerance for unsafe practices?
- Do all of our executives and managers act as role models and promote this safety culture?
- Do all employees believe in the safety culture?
- Are safety procedures followed all the time?
- Is there a trusting relationship between management and labor?
- Do the employees believe that management responds ethically regarding safety issues?

IV. EFFECTIVE COMMUNICATION

A. TOTAL QUALITY COMMUNICATION

Effective communication is the effective interchange of information and relationships for TQM (or safety). In TQM organizations, information flows are free, direct, and continually reinforce the product and customer-oriented, fact-oriented, and win-win mindset.

According to Creech,[22] the following ideas help to make up and down communication boundless, honest, and unstilted.

- All communications should be in the language of trust. Implying a mistrusting atmosphere will impede the flow of factual and emotional information necessary to understand situations, solve problems, and make effective decisions.
- All communication should be in simple and direct language. Using difficult, abstract and convoluted language prevents the clear transmission of ideas.
- Everyone should be open to dissenting views and acceptance of bad news. Stifling dissention robs decision-makers of the base of ideas and factual information necessary for analysis.
- Communicators should verify that the proper message was received. Asking others if they understood ensures that the communications are getting through.

Similar ideas are put forth by Petersen who cites several conditions that encourage upward communication.[23] These conditions are:

- genuine two-way communication at all levels of management, including the very highest;
- accessibility and responsiveness of supervisors;
- welcoming new ideas;
- providing visible rewards for new ideas;
- accepting criticism; and
- being sensitive to the employees.

In summarizing his analysis of TQM communication, Creech provides a specific recipe for effectiveness.[24] "Give high priority and pay great attention to the communication flow. On key issues augment the hierarchical flow. Go several layers deep. Talk numbers as well as words. Ensure full comprehension throughout. Replace all inhibitions to upward communication with full openness. Provide the requisite means and adequate incentives to make it work. Listening, hearing, and caring are the catalysts which make it (communication) thrive."

Effective communications for change and improvement should focus on the front-line supervisor. Larkin and Larkin[25] report that front-line supervisors should introduce changes to employees in face-to-face discussions that focus strongly on facts. They see this communication path as the best way to get the important change messages to employees, recommending that fully 80% of management's communication efforts should be targeted to the front-line supervisors. Specifically they suggest giving the supervisors briefing cards and forms for soliciting employee advice.

B. SAFETY COMMUNICATIONS

For safety, communication should be effective and innovative including using a wide variety of techniques such as: training, meetings, memos and postings, and informal discussions, to get the safety messages out and understood. All safety communications should reinforce the safety culture, vision, objectives, and commitment, and the expectation that the workers are responsible for safety. There should be a high priority and attention evidenced in safety communications. The safety message should be communicated clearly and consistently. There should not be any inhibitions to upward communication with full openness. There should not be any record of reprisal or unofficial harassment against workers reporting safety problems. Management should positively respond to safety reports, and employees should hear about actions to address any problems identified.

For effective communication, everyone should understand and use a common language. For example, using the term 'defect' is a way of referring to errors, failures and breakages without placing blame.

Feedback to employees is an important part of safety communication. Geller[26] offers specific guidance for effective behavior-changing safety feedback. His guidelines include: deliver the appropriate feedback immediately after the behavior, give specific direction for improving the unsafe behavior, use objective and straightforward language, don't mix messages, give corrective feedback just prior to the next opportunity, and tailor the feedback to the person and situation.

EFFECTIVE COMMUNICATION: ASSESSMENT QUESTIONS

- Do we use a variety of effective communications regarding safety including:
  - routine training?
  - meetings?
  - memos?
  - newsletters and postings?
  - informal discussions?
- In all of our communications is there reinforcement of the safety culture, vision, objectives, and commitment?
- Do all our employees freely discuss safety problems and offer safety suggestions to their supervisors and managers?
- Does management positively and quickly respond to safety concerns and suggestions?
- Are employees protected from reprisals and harassment?
- Do managers and supervisors provide effective safety feedback to employees?

V. ORGANIZATIONAL AND EMPLOYEE KNOWLEDGE

Knowledge enables the organization and the employees to achieve and to continuously improve. The organization should strive to achieve states of adaptive and generative learning, and the workforce should understand TQM (or safety) concepts and have the skills to perform activities correctly. Adaptive learning helps accommodate changes (is flexible and responsive), and generative learning is the ability to expand capabilities and generate something new.

As cited by Roughton,[27] organizational knowledge also includes the understanding that "unsafe acts and conditions are the critical common pathway from which injuries arise. Thus the ability to control behavior distinguishes those organizations with low accident frequencies."

According to Brown et al.,[28] the following are key characteristics of a learning organization:

- alignment of all organizational members to a shared work safety vision;
- learning is spontaneous;
- experimentation and risk taking are recognized as integral to learning;
- management models learning;
- group learning is emphasized over individual achievement.

The steps necessary to create a learning organization are described by Brown et al.[29] as:

- increase the monies spent on work safety education and training;
- organizational (create learning infrastructures, promote experimentation, empower employees);
- leadership (champion a shared vision, manage confusion and tension, model learning);
- team (practice art of dialogue, develop habit of reflection, embrace change).

In terms of safety, the organization would need to demonstrate that the organization and the workforce exhibit the knowledge and characteristics necessary to achieve work safety objectives. The organization should demonstrate the ability to generate and use new ideas for safety management, and should also demonstrate the ability to adapt ideas and techniques for others to enhance safety. It should be clear that all workers and managers understand the quality and work safety programs and their core principles and have the skills to work safely. Also, the organization and workforce should understand that unsafe acts and conditions are the critical common pathway from which injuries arise and that the ability to control behavior distinguishes those organizations with low accident frequencies.

In TQM, deploying the technical, business and organizational activities throughout the company promotes learning. Management and professional staff teach the workforce about technology, business intelligence, research, quality functions, and policy making.[30] The managers and technical staff tend to become deprofessionalized and knowledge is spread out, not hoarded.[31] In addition, companies should set minimum levels of skills for employees and then continually improve those levels. Such minimum skills include meeting skills, using E-mail, and planning and implementing.

Solomon[32] contrasts organizational learning with traditional learning.

- Training goals are based on the corporate strategy and user needs rather than user requests.
- Learning focuses on core competencies and long-term strategic plans rather than immediate needs and short-term plans.
- Training is delivered in real time, on request rather than periodically scheduled.
- The education approach is to design learning experiences rather than deliver knowledge.
- Education is self directed, designed with the participants rather than instructor driven.
- Employees receive training support and lifelong development rather than just skills.
- Needs assessments are done jointly by individuals and trainers rather than by trainers.

ORGANIZATIONAL AND EMPLOYEE KNOWLEDGE: ASSESSMENT QUESTIONS

- Does our organization and all of our workforce exhibit the kinds of knowledge and characteristics necessary to achieve work safety objectives?
- Do we generate and use new ideas for safety management?
- Do we adapt ideas and techniques from others to enhance safety?
- Do all our people understand the quality and work safety programs and their core principles and do they have all the skills necessary to work safely?
- Do our people strive to learn more and develop their skills?
- Do employees participate in the training program other than as students?
- Is there a minimum safety skills and knowledge level set for employees?

28

VI. EMPLOYEE EMPOWERMENT

Employee empowerment means that the workforce is enabled to fully participate in making TQM (or total quality safety) a reality. As defined by Biech,[33] "empowerment is a set of conditions that must be created that allows people to reach their maximum potential and frees them to act in the most beneficial way for the customer, their department and the company." Empowerment is the authority to act independently to meet expectations. It is exercise of responsible freedom in the customers' behalf.[34]

Empowerment improves employee productivity, involvement and ownership; it opens avenues of change in the business culture and improves operations; it fosters communication at all levels; it makes best use of employees' talents and knowledge; it creates careers over jobs and establishes a basis for trust; it produces action over reaction and results over excuses; and it builds the opportunity for continuing improvement by installing a network of problem solvers throughout the organization. Empowered employees enrich themselves and their company through personal achievement, growth, and customer-directed problem solving.[35]

Empowerment involves giving employees the authority, ability, and resources to analyze and attempt to solve problems or satisfy customers on their own. To empower employees the organization should first determine what organizational factors prevent it from occurring and then what must be done to make the employees believe they are empowered. Rules and regulations that limit the employees' authority and ability and mandate unnecessary oversight must be rethought and revised.

According to Hradesky,[36] empowerment requires three acts: the act of empowerment; the act of process empowerment and the act of taking responsibility. Hradesky defines these acts as follows:

- The act of empowerment means: giving expectations in terms of objectives, goals and deadlines; giving guidelines in terms of policies and procedures; giving authority and specifying the boundaries of decisions; giving adequate resources; and giving the skills and techniques needed to achieve.
- The act of process empowerment means: determining the means to measure performance; providing accountability for teams and individuals; reviewing the status or progress; giving periodic feedback; and providing appropriate positive or negative consequences.
- The act of taking responsibility means: acknowledging, agreeing, and accepting the empowerment and then acting responsibly.

For safety empowerment, Hradesky's requirements become:

- The act of safety empowerment means: giving employees specific safety expectations; giving safety policies and procedures; giving the authority to make work safety decisions, such as stop work; giving adequate safety resources such as personal protective equipment; and giving sufficient safety training to work safely and fully understand the safety programs.
- The act of safety process empowerment means: giving ways to measure safety performance; making sure all employees, teams, supervisors, and managers are accountable for safety; reviewing safety performance status through observations, inspections, audits and reviews; giving employees periodic feedback on safety performance; and providing appropriate consequences.
- The act of taking responsibility for safety means: acknowledging, agreeing, and accepting the safety and empowerment expectations and means and then acting responsibly.

In slightly different terms, Biech[37] discusses the key elements of empowerment. These are: educating, enabling, and encouraging.

- Educating is: understanding people and needs; mentoring, coaching, counseling, and teaching; delegating challenging tasks for development; modeling correct behaviors;

providing opportunities to learn new skills; and providing timely and adequate information flow to do the job.

- Enabling is: providing resources to get the job done; removing roadblocks and driving out fear; sharing accountability; allowing experimentation and risk taking; and delegating authority and responsibility.
- Encouraging is: expecting the best from people; trusting and encouraging innovation, ideas, and risk; and giving meaningful work that provides challenge and variety; rewarding and reinforcing accomplishments; and soliciting feedback.

In Biech's terms, employees are educated, enabled, and encouraged for safety empowerment. They have appropriate role models and are given the resources and information they need to work safely. These include feedback and coaching if unsafe actions are observed. They are not afraid to act and report on safety issues. Everyone is expected to exhibit correct behaviors and all employees are trusted to work safely and follow safety rules and procedures. All safety accomplishments are reinforced through appropriate recognition and reward systems, and all employees are interested in how they are doing.

Individuals who are empowered:
- Take personal responsibility,
- Continually assess their own performance,
- Manage their own performance, taking corrective action when necessary,
- Seek help when they need additional resources, and
- Take the initiative to help others in other areas.

Hayes[38] describes how the extent of empowerment can be determined by a set of employee beliefs, which I have rephrased in terms of safety:
- I can do anything to do a safe job.
- I have the authority to correct safety problems which occur.
- I can be creative when dealing with problems.
- I don't have to go through a lot of red tape to change things.
- I have a lot of control over how I do my job.
- I don't need management's approval before I handle safety problems.
- I am encouraged to handle job-related safety problems by myself.
- I can make safety-related changes on my job whenever I want.

Empowerment is also closely related to the following job variables: participation in decision making; organization-based self esteem; task autonomy, variety, and importance; job satisfaction; and the supervisor's commitment to safety and quality.

In a safety-empowered organization, employees should:
- Inspect for hazards and recommend corrections or controls,
- Analyze jobs to locate potential hazards and develop safe work procedures,
- Develop or revise general rules for safe work,
- Train newly hired employees in safe work procedures and rules,
- Train co-workers in newly revised safe work procedures,
- Provide programs and presentations for safety meetings, and
- Assist in accident investigations.

EMPLOYEE EMPOWERMENT: ASSESSMENT QUESTIONS

- Are all our employees prepared for and empowered to fully participate in work planning and safety, including making in-field safety decisions?
- Do all our employees have the authority, given by management, to act independently to meet safety expectations?
- Are all employees given expectations, guidelines, resources and skills necessary to act independently?
- Do employees:
  - inspect for hazards?
  - analyze jobs?
  - develop safe work practices?
  - train new employees in safety?
  - assist in accident investigations?
- Do employees have stop work authority if unsafe conditions develop?

VII. EMPLOYEE RESPONSIBILITY AND EXCELLENCE

Employee responsibility means that the workforce is prepared for, accepting of, and has taken ownership of the processes for total quality management (or for work safety).

A. RESPONSIBILITY

Taking responsibility is crucial for full and effective empowerment. For work safety it means: acknowledgment of work safety and all its objectives; agreement with work safety policies and programs; acceptance of roles in the organization and acting responsibly as a safe worker/employee.

In accepting responsibility for a safe work environment, workers would:
- cooperate with peers and supervisors in working safely and learning from safety coaching;
- involve themselves in safety meetings by actively participating and contributing to discussions;
- be receptive to safety issues; and
- voice any safety concerns promptly.

In addition, employees would make sure that co-workers would also provide peer support, would be proactive regarding safety concerns, and would hone their work and behavioral observation skills.[39]

B. EXCELLENCE

Excellence means that the employees strive to improve, work correctly and safely, and truly believe in doing the right things right. Excellence in safety means that employees follow procedures, utilize proper equipment and protective gear, understand and follow safe work behaviors, contribute to continuous improvement activities, communicate with others and generally promote the safety culture.

In addition, employees would:
- learn and use quality and analysis tools for safety;
- learn to work and contribute in a mutually supportive team environment;
- understand and follow all safety policies, rules and requirements;
- learn to participate in audits, inspections, reviews, and incident analyses; and
- learn to observe and analyze the work of others and coach for improvement.

Harrington[40] cites that employees working toward excellence also must:
- understand what their performance levels are;
- be willing to take risks;
- be willing to learn new ways of doing things;
- be willing to take on new jobs;
- think creatively,
- be uncomfortable with the status quo, and
- be willing to share credit with others.

Individual excellence results from pride, self-esteem, and dedication. In an organization which promotes individual excellence:
- employees are expected to develop improvement ideas and contribute to the testing of processes and procedures;
- rewards and recognition systems promote creativity and risk-taking in the employee's contributions to safety improvement efforts;
- career development is emphasized by rewarding learning new skills and assuming increased safety responsibilities;
- management is open to and accepting of new ideas; and
- all safety improvements, both big and small, are important and are recognized and communicated to the organization.[41]

EMPLOYEE RESPONSIBILITY AND EXCELLENCE: ASSESSMENT QUESTIONS

- Are all employees prepared for and sharing responsibility for a safe work environment?
- Do all the employees:
  - acknowledge work safety and all its objectives?
  - agree with work safety policies and programs?
  - accept their roles in the organization and act responsibly as safe employees?
- Do I see all the employees working safely all the time?
- Do employees: participate in audits? participate in inspections? participate in incident analyses?
- Do employees analyze the safety performance of co-workers and coach them for improvement?
- Do employees learn and use quality tools for safety?
- Are employees expected to contribute safety improvement ideas?

VIII. MANAGEMENT BY FACT

As described in the Malcolm Baldrige National Award Criteria,[42] modern business management is built on a framework of information, measurement, data, and analysis. Measurements derive from the company's strategy and cover all key processes and the outputs and results of these processes. Facts and data are needed for performance improvement and assessment. Performance indicators which are measurable characteristics of the products, services, processes, and operations are used to track and improve performance. These indicators should be the best representations of the factors leading to improvement in customer, operational, and financial performance. What gets measured gets managed. Bottom line measures are after the fact, and solid service/quality measures are an early warning system of what the bottom line will be.

Creech[43] explains that measurement of specific goals provides important results for an organization.

- It helps to focus on what is important.
- It provides objectivity.
- It helps to recognize what needs to be fixed.
- It guides improvement of the right things.
- It provides motivation to improve the score.
- It aids the process of decentralization, by helping to maintain cohesion and control.

Measurements also increase commitment when tied to the empowerment and reward processes.

Performance measures should be simple and easy to use, provide fast feedback and should foster improvement rather than being used just for monitoring. They should be specific, timely, achievable, and realistic.[44]

In setting measurements and standards, companies tend to make the following types of mistakes.

- They use measurements as reasons to punish.
- They act too slowly in providing feedback.
- They have secret or selective feedback.
- They don't react to problems.
- They have measurements that are set in stone.[45]

Other mistakes include:

- compiling too much data;
- failing to base decisions on data; and
- having unspoken, incomplete, or inconsistent measurements.[46]

In any case, data should be used to control the quality system, to furnish information to management, and to foster improvement through analysis.

Results and achievements are emphasized by Brown et al.[47] who discuss the problems with TQM measurement systems. They point out that organizations sometimes focus on behaviors rather than accomplishments; they emphasize courtesy rather than competence; they focus on internal improvements while neglecting external needs; and they fail to identify key process variables.

A. SPECIFIC TQM MEASURES

Exactly what measures are used to assess TQM performance. In a recent survey conducted for The Conference Board, the following measures were found in TQM-related performance appraisals.[48]

- demonstration of quality leadership
- formal measures of external customer satisfaction
- formal measures of internal customer satisfaction
- employee involvement in establishing quality goals
- measures of business processes
- measures of cycle time
- measures of defect/errors in products or services
- measures of employee satisfaction and morale
- measures of involvement external to the company
- measures of quality related behavioral competencies (initiative, teamwork, coaching, and empowerment)
- measures of skill and knowledge acquisition
- team/group measures of performance
- use of peer evaluations

This list is similar to that outlined by Clemmer[49] as important for managing quality. He cited:
- customer satisfaction measures;
- process measures to form a basis for continuous process improvement;
- product measures or quantifiable measures of reliability, conformance to specifications, or effectiveness of the basic product;
- project measures for project goal achievement;
- cost of quality and cost of bad quality; measuring the quality improvement process; and
- benchmarking.

Specifically, Brown et al.[50] show how common measurement mistakes can be corrected.
- Instead of "training hours" use "percent of training plan objectives met."
- Instead of "total training expenses" use "return on training investment."
- Instead of "number of suggestions" use "number of suggestions implemented."
- Instead of "number of teams" use "team-based performance improvements."
- Instead of "number of quality awards" use "employee satisfaction with award program."
- Instead of "number of executive quality speeches" use "employee survey on executive commitment."

B. SAFETY MANAGEMENT MEASURES

In safety management, typical upstream indicators are safety and health related behavior, practices and conditions, and downstream indicators are the actions taken by management to positively change the safety and health system. Crucial to using these indicators is development of a list of critical behaviors based on previous injuries and measurements of conformance to safety requirements.[51]

Krause[52] discusses the use of accident rates as safety measures, observing that they must be used with caution. He notes that since normal accident (or injury) rates are low, apparent rate variations may often be statistically invalid. Specifically he provides an example where a quarterly change from 4.0 to 0.0 injuries for a 100-man group that averages 2.0 injuries per year has no statistical significance. Krause[53] also notes that using accident data as a primary safety performance indicator diminishes credibility in the entire safety program. It does this by being inactive rather than proactive, by causing management to often overact to events, and by subverting incentive programs to cause accident underreporting. Misclassifying accidents is also a common way of subverting performance appraisal systems based on accident results.[54]

Geller[55] states that safety should be measured and displayed to the entire workforce in terms of "achievements" with emphasis on processes that decrease the injury rates. Achievements can be environmental (safer equipment, correction of hazards, etc.), personal (training, celebrations, additional safety personnel), or behavioral (observations and trending of work practices).

An important safety management measure is the real or total cost of accidents and injuries. According to Grimaldi and Simonds[56] the total cost of accidents is the sum of direct and indirect costs. The direct cost is just the insured dollar loss total. The indirect costs comprise a long list of factors including:
- wages for working time lost by uninjured workers;
- net cost to repair, replace, or straighten up material or equipment damaged in the accident;
- uninsured wages paid for working time lost by injured workers;
- extra costs due to accident-related overtime work;
- supervisor's wages for accident-related activities;
- wage cost of decreased injured worker output after return to work;
- cost-of-learning period of new workers;
- uninsured medical costs;
- accident investigation costs incurred by managers and others;

- clerical costs for processing compensation or insurance forms; and
- miscellaneous unusual costs, such as equipment rentals or hiring new workers.

Johnson and Burke[57] list various estimates of accident and fatality costs. OSHA estimates the average lost-time injury costs employers $4,000. The National Safety Council estimates the cost of a disabling injury to be $29,000. Total fatality cost estimates range from $790,000 (OSHA) to $8 million (EPA).

MANAGEMENT BY FACT: ASSESSMENT QUESTIONS

- Is our work safety management system based on specific, timely, achievable, and realistic performance measures; on positive performance indicators; and on a framework of information, measurement, data, and analysis?
- Do all of our executives review safety results?
- Are measures other than accident statistics used as the primary performance measures?
- Are accident statistics used to assess performance in a statistically valid way?
- Does the workforce understand what statistical analyses are valid?
- Do our performance measures include:
    - employee satisfaction and morale?
    - safety-related behavioral competencies?
    - safety skill and knowledge acquisition?
    - safety suggestions implemented?
- Do we measure the real costs of accidents and injuries?
- Are safety achievements measured and displayed to the workforce?

IX. LONG-RANGE VIEWPOINT

A. TOTAL QUALITY MEANING

One of the Malcolm Baldrige National Quality Award Criteria[58] core values and concepts is long-range view of the future. This means there is a strong future orientation and willingness to make long-term commitments to all stakeholders. In planning for the future, organizations need to account for and anticipate the many types of changes and events that might affect customer expectations. These changes and events include: technological developments, changing regulatory requirements, and changes in societal expectations.

In a global sense, focusing on short-term goals is said to shorten imagination and vision. In striving to improve, personal viewpoints may need to be extended toward the attainment of broader, more visionary goals.

Too many organizations are oriented to the present instead of the future.[59] The reasons for the present orientation are:

- Short-term pressure. There is too much pressure for short-term results. This is evident in magazines and papers in discussions of changes in quarterly profits, sales, or in daily fluctuations of stock prices. This pressure results in short-term activity to address the symptoms of problems rather than the causes.
- Organizational design. Organizations are designed for stability and control rather than adaptability and change. Bureaucratic, hierarchal organizations still predominate and are slow to make changes.
- No one responsible. In most organizations, the planning function is too directed to operational planning rather than strategic planning. The focus is on the short-term.

Narus[60] also points out that future activities and thinking should be spread out over two horizons. The first horizon is that of commitment. This planning horizon is usually three to five years ahead. This is the time that reasonable resource, design, and implementation commitments can be made and relied on. The second horizon is the screening horizon. This is the five to twenty year time period that the company needs to understand to make intelligent strategic decisions. This is the period when the fruits of commitment decisions are realized.

B. SAFETY IMPLICATIONS

A strong long-range viewpoint of the future is demonstrated by the company's plans, strategies, and resource allocations. These should show a clear commitment to long-term planning. A strong commitment to planning for the future also would be evident in employee development activities where employees would be educated so that they can function and contribute to the organization in the changing environment. The long-range viewpoint extends to viewing the employees as lifetime workers.

In terms of safety there should be a strong future orientation and planning for safety and employee development with a willingness to make long-term work safety commitments. In striving to keep abreast of changes in technological developments, the organization should have a strong technical base involved both with the safety of core processes and activities, and with changes in safety technology, products, and services. This implies that there is a strong measure of professionalism and technical ability among the technical staff with an ongoing commitment to training and development.

In assessing the organization's long-term commitment to work safety, you would want to see evidence of that commitment and dedication with continuing safety education and training, and safety improvements planned and budgeted. Other assessment factors would include whether there is an understanding of what changes in safety regulations and requirements were coming and a demonstration that planning was underway to evaluate those changes and accommodate them.

LONG-RANGE VIEWPOINT: ASSESSMENT QUESTIONS

- Is there a demonstrated strong future orientation and long-range planning for safety and employee development with a willingness to make long-term work safety commitments?
- Do we have a strategic plan for safety development?
- Do we plan and commit for future employee development and for improvements in workplace safety?
- Do we keep aware of regulatory changes and initiatives and do we plan for them?
- Does our budget reflect the continuing commitment of money and resources for training and safety?
- Does the company avoid short-term reactive approaches to accidents, incidents, and safety problems in favor of longer-term corrective actions?

X. STATISTICAL PROCESS CONTROL

Statistical process control (SPC) is the use of statistical methods to control and reduce variation around requirements for all company quality/safety activities and processes. The goals of SPC are: conformance to specification, removal of unacceptable variation, and control within established tolerances.[61] In SPC, work outputs and results are recorded graphically and once trends are identified, either the tolerances can be made more stringent or adjustments can be made to the work

36

process.   For safety management purposes, the goal of using SPC would be to establish a controlled, accident-free environment. The process would involve graphing and then analyzing important safety-related information such as incident rates, safety behaviors, or implemented suggestions.

SPC (according to Deming) is best used to guide the understanding of the causes of accidents and is not to be relied upon for total control. For accident analysis, this means that SPC is best used for relatively frequent accidents where statistically valid data are available. SPC is just not applicable for control of infrequent catastrophic accidents where random variation obscures the statistics. Analysis and control of infrequent accidents are functions of prior anticipation and effective design.

A. STATISTICAL CONTROL TOOLS

The main types of charts and diagrams used for statistical analysis and control are the:

- **Pareto** chart which shows the relative importance of data on bar graphs which descend in importance from left to right. Typically, Pareto charts are used to rank problems or causes so that priorities for analysis or action can be determined. In the simple Pareto chart shown in Figure 4, the first type of defect is the largest cause of equipment defects and should be the top priority for a corrective/preventive analysis.
- **Run**   (trend) chart which shows the data or results of a process plotted over a period of time. Run charts are used to identify trends and changes. The simple run in Figure 5 shows a plot of injuries per month. Once a median is constructed, run chart data can be checked for stability and checked for patterns (trends up or down, points positive or negative, and points alternating up and down).

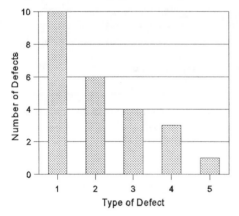

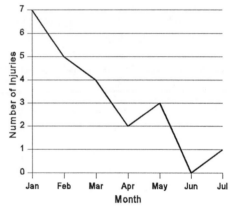

Figure 4. A sample pareto chart showing the significance of the causes of defects in a component.

Figure 5. A sample run chart showing the downward trend in injuries for the first part of the year.

- **Histogram**   which identifies and displays the distribution of data on bar graphs, typically to show the frequency with which something occurs. The simple histogram in Figure 6 below shows the variation in the number of defects per week in a production run. The distribution of data shows a normal (not skewed) variation.
- **Scatter diagram** (XY graph) which shows relationships between two variables. In the simple scatter diagram in Figure 7, there is a good relationship between the weight and the height of pieces of metal.

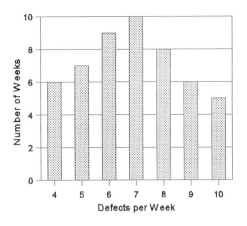

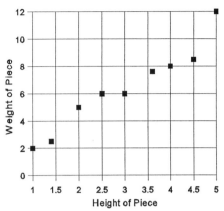

Figure 6. A sample histogram showing the changing number of defects per week for a product.

Figure 7. A sample scatter diagram showing the relationship between height and weight of an object.

- **Tally chart** (check sheet, data sheet) which is a primary tool for data gathering. The tally chart simply lists the numbers of hits or items. Tally charts can be used to record safety inspection results or to record behavioral observations. When developed, tally charts should be made easy to use as shown in Figure 8.
- **Control chart** which shows the variation within a system. With statistically determined upper/lower limits, control charts determine if variations are normal or not and measure the impact of changes. Control charts are the primary means of controlling processes. With the upper and lower control limits set at three standard deviations, 99.7% of the charted data should fall within the control range. Abnormal data outside the range are attributed to special cause variation. A typical control chart is shown in Figure 9.

Problem	Mon	Tue	Wed	Th	Fri
Jam	///	//	////	//	///
No Paper	//	/	//	/	/
Blank Page	/		/		/
Torn Paper	//	/	//	/	

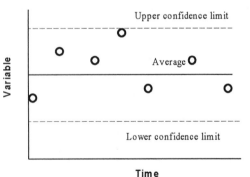

Figure 8. A sample tally chart showing the causes of problems with an office printer during a week.

Figure 9. A typical control chart showing upper and lower confidence limits around a mean level.

Shearer[62] describes different types of control charts that are used to help analyze and control processes depending on the type of data (either count, attribute, or variable) obtained, and Krause[63] describes the application of control charts to occupational safety data. Count data are data which give actual occurrences, with no time or area boundaries - such as defects on a product. Attribute data list items in a finite group, such as the number of cars with defective wheels. Variable data are data measured during a process. Variable data can be subdivided whereas count and attribute data are integral.

Among these other control charts are:

- c-charts which are used for count data with constant sample sizes (such as injuries per year).
- u-charts which are used for count data with variable sample groups (such as rework hours per project).
- p-charts which are used for attribute data (such as the fraction of a group showing poor behaviors).
- X-bar or R charts used for continuously monitored variable data.
  - The X-bar (or average) charts are used to monitor the shifts in a process centerline.
  - The R (or range) charts monitor the process range or dispersion.
- X and moving Range Charts which are used with variable data to continuously monitor a process.
  - The X charts plot raw observed data.
  - The Moving Range Charts plot point to point differences in the data.

B. APPLICATION TO SAFETY

A direct application of SPC to safety management is the measurement and statistical analysis of upstream behavioral indicators.

Krause[64] presents five principles that are important in doing a valid statistical analysis of accident data. These principles are:

1. Carefully define the terms. Define what is an accident. Determine what time period is involved. Specificity in defining terms is important for obtaining a clean data set for analysis.

2. Analyze rates of occurrence - the number per exposure hour. Analyzing a rate rather than a total keeps the data sets uniform and permits valid statistical analysis to be accomplished.

3. Check the accident frequency rate for all relevant variables - such as age, years on job, shift, gender, time of day, department, function, etc.

4. Cross atypical variables with the standard variables - atypical variables are things like parts-of-the-body, that are not exposure-hour related. Standard variables are things like department or shift. This could help determine what the proximate accident causes are.

5. Use statistical techniques to determine significance - for example, use the chi-square test at the 5% significance level. This says that the observed variation has only a 5% chance of being random and the analysis is probably determining something of significance.

STATISTICAL PROCESS CONTROL: ASSESSMENT QUESTIONS

- Are statistical process analyses and methods used to manage and control workplace safety in all departments and processes?
- Are accident and injury statistics analyzed in a statistically valid way?
- Are any of the following charts or diagrams used for statistical measurement and control: pareto charts? run charts? scatter diagrams? histograms? control charts? or tally charts?
- Do the managers and employees know what these charts/diagrams are and how to use them?
- Does everyone have a good understanding of accident statistics and what results are valid?

XI. STRUCTURAL PROBLEM SOLVING

Structural problem solving means the use of effective techniques to solve problems, prevent future problems, and create opportunities for improvement. It is an integral part of the TQM improvement process. For safety, structural problem solving means that effective problem analysis and problem solving techniques are used to analyze any accidents and safety problems and arrive at effective corrective and preventative actions. The range of accidents analyzed should include near-misses, since their number is much more frequent than actual accidents.[65]

A. BASIC PROBLEM SOLVING STEPS

Problem solving involves the following basic steps:

- **Defining the problem:** consider the symptoms, the history and trends. Include areas of marginal performance and any new requirements.
- **Collecting data:** use surveys and observations, collect industry data, conduct company interviews, and evaluate historical trends and data.
- **Analyzing data:** use typical analysis tools such as Pareto diagrams, the null-op method, etc. Typical analysis tools and techniques are listed below.
- **Developing alternatives:** trends and data patterns lead to possible solutions. The alternatives could be to improve, increase or revise, or to decrease or hold fast.
- **Evaluating alternatives:** use feasibility analysis, pilot programs, financial analysis and predictions, market analysis and potential, criteria development and utilization.
- **Selecting the optimum alternative:** through the application of selection criteria, financial impact, feasibility, market potential and projections, and company fit - use the screening matrix technique to analyze and decide or use the weighted factors approach.
- **Implementing:** establish project tasks and schedules, and the required manpower and financial resources. Evaluate organizational considerations, project responsibility requirements, progress reporting, and monitoring systems. The how-how diagram (a modification of the why-why diagram described below) can be used to determine implementation steps.

B. PROBLEM SOLVING TOOLS AND TECHNIQUES

There are many tools and techniques which can be used to analyze problems and develop appropriate responses. A non-exhaustive list of typical analysis tools and techniques includes:

- **Pareto Diagrams** - specialized bar graphs showing the relative frequency of events or the relative importance of problems. The basis is that 80% of the problems result from 20% of the causes. Pareto diagrams help to focus on the principal aspects of a problem and decide on the most effective solutions.
- **Pareto Cause Analysis** - in which cascading Pareto diagrams are used to delve deeply into the frequencies of causes. The technique evaluates subcauses to identify the root causes. Typically the top one or two causes identified in the first analysis are themselves analyzed in the next layer(s) until a root cause is identified.[66]
- **Delphi Method** - in which a questionnaire is sent to experts for consideration and response. Upon receipt of the responses the questionnaire is revised or summarized and resubmitted. The process continues until a consensus is achieved. In this time-consuming, yet effective method, the separate inputs of experts are considered on their individual merits.
- **Null-op Method** - a technique where a group determines the factors that are important for making a decision and assigns weights to them for choosing among alternatives.
- **Deductive/Inductive Reasoning** - the processes of logical reasoning and grouping to draw conclusions. Deductive reasoning involves drawing conclusions from premises, usually in the form of a syllogism. Inductive reasoning involves creatively grouping related information and classifying them.

- **Nominal Group Technique** - a small group process for generating and ranking ideas. In this process, ideas are generated, recorded, clarified, and voted upon. Usually the voting takes place iteratively by secret ballot, with the ideas garnering the fewest votes tossed out after each round of voting. This process diminishes the effect of a dominant person, allowing all participants to contribute more equally.

- **Fishbone Diagram** (or cause and effect diagram) - used to identify all possible causes of a problem by brainstorming and progressive clarification. Using this method helps to prevent quick judgements and subjectivity in an analysis. As shown in Figure 10, first a main line is drawn horizontally, then the proximate causes are depicted as the main branches off the line, and with more analysis, contributing causes are depicted as stems off each branch. Once the problem is thoroughly analyzed, possible solutions can be developed.

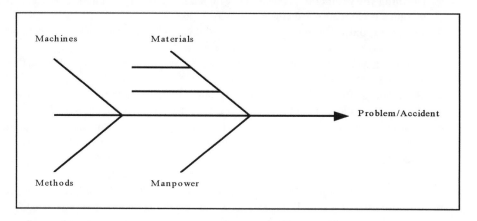

Figure 10. An example of a fishbone diagram showing lines for proximate causes and contributing causes.

- **Why-Why Diagram** - a form of tree diagram that allows possible causes of a problem to be identified systematically. Starting with a problem written down, an analysis tree is built up by continually asking, "Why?" The first "Why" results in listing the main causes, the second and third "Whys" result in identifying the contributing factors.

- **Affinity Diagrams** - group development of ideas and issues followed by their arrangement into affinity (or similarity) families. The ideas can be generated by brainstorming. Once all the ideas are arranged, headers are created to describe the affinity families. This process is good for large, complex issues with little structure.

- **Brainstorming** - free unrefined thinking to generate creative ideas without judgement. It is used to identify problems; collect improvement ideas; set directions; clarify inputs, outputs, and activities; identify root causes; suggest creative solutions; and identify pros and cons.[67] The goal of brainstorming is to get a quantity of ideas with the participants not commenting on or analyzing them

- **Screening Matrix** - for analyzing and deciding among alternatives. A simple (3 by 3 or 4 by 4) matrix is constructed with evaluation criteria denoted on the axes. Individual ideas are analyzed and categorized in the matrix.

- **Comparison Matrix** - for pairwise comparison of alternatives. A matrix is established listing all alternatives horizontally and all alternatives-minus-one vertically. Each alternative is compared against all the others, and the ones with the higher scores are selected for more detailed analysis.

- **Boundary Analysis** - used to explore the parameters of problems. Typically used for problems that are not large, not found everywhere, and not always found. The boundaries of the problem are defined through a questioning process and the answers analyzed to search for clues to the solution.
- **Force Field Analysis** - for identifying and analyzing the driving and restraining forces regarding a situation. Pictorially, as shown in Figure 11, the driving and restraining forces of a problem are shown as opposing arrows. The forces can be weighted in importance or ranked in strength. The analysis considers how to improve the positive forces and restrain or remove the negative forces.

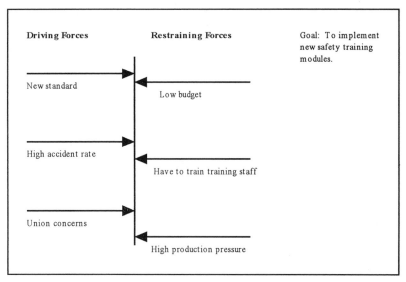

Figure 11. A sample force field analysis for implementing new training program modules.

- **Quality Function Deployment** - a chart for converting requirements into implementation techniques. Requirements are listed (vertically), with corresponding technical requirements or how they are met listed (horizontally) with weighted rankings used to indicate importance.

C. VOTING TECHNIQUES

Voting techniques are used to choose among alternatives. Typical voting techniques include:[68]

- **Multivoting:** used for making a large number of choices smaller. In this process, each member of a group votes for 1/4 - 1/3 of the possible choices. The choices with the highest scores are retained, the others discarded.
- **Nominal Group Technique:** where team members choose the best 3-5 items and rank the individual choices. Usually combined with silent brainstorming, discussion, and prioritizing.
- **Pairwise Ranking:** used for small groups of less than 10 items. Each item is ranked against all the others on the list by team or individually. A comparison matrix can be used to display the results of the ranking.
- **Ten-Four Voting:** used for lists of from 10 to 15 items. In this process each group member has 10 votes to apportion among the items with up to 4 votes allowed for any single item.
- **Priority Voting:** used to rapidly reduce lists of items. Each member votes for as many items as wanted. Items getting 30% or more of the votes are retained. In subsequent rounds, the participants vote for 50% of the items. The voting continues until only 2 to 6 items remain.

STRUCTURAL PROBLEM SOLVING: ASSESSMENT QUESTIONS

- Are effective analysis and problem solving techniques used to identify problems, and analyze accidents and safety problems?
- What specific problem analysis and solution techniques have personnel been trained to use?
- Are injury and illness trends analyzed, so that patterns with common causes can be identified and prevented?
- Are there documented examples showing how the problem solving techniques have been used?
- Have the employees been trained in the basic steps of problem solving?
- Are any of the listed voting techniques used to decide among alternative courses of action?

XII. THE BEST TECHNIQUES

Best techniques means the use of world-class techniques for processes and procedures to prevent problems. For safety world-class safety techniques from other organizations are identified and applied to the particular situation (industry, process, hazards, etc.). The best techniques are determined by the processes of **baselining** and **benchmarking**.

A. BASELINES

Baselines are critical components of the improvement process. They provide the points of reference for all improvement efforts based on managements communication of customer expectations tied to measurements and standards of excellence.

For safety, baseline surveys are recommended by OSHA's Voluntary Protection Program Guidelines[69] (see Chapter 4). The intent of safety baselines is to develop a comprehensive picture of safety conditions and problems which then serves as a starting point for improvement efforts and follow-up routine inspections. After the baselines, additional comprehensive assessments may be conducted to analyze for less obvious problems or to review areas which have given rise to incidents.

OSHA recommends that personnel conducting the comprehensive surveys should be appropriately trained and qualified. In specific, since these surveys are beyond the scope of normal routine inspections, certified professionals are recommended by the Department of Energy.[70]

B. BENCHMARKS

Benchmarks are comparisons and analyses against others. As defined by Spendolini, benchmarking is a continuous, systematic process for evaluating the products, services, and work processes of organizations that are recognized as representing the best practices for the purpose of organizational improvement. Benchmarking should not be used only once, to search for quick and easy solutions to problems, but it should be used for strategic planning, to find new ideas, to compare products and process, and to set goals.[71]

Benchmarking is an important function because it can provide the authority for making changes. Benchmarking is also an indicator of the degree of organizational willpower to understand and carry out the steps needed to succeed.[72] When better techniques are observed, copying them is a vital, valid learning technique. Copying fosters listening to others, breaking local dogmas, modesty, clarity of inadequacies, learning the systems of others, and the habit of study-evaluation-and adaptation.

There are various types of benchmarks. Benchmarks can be internal, competitive, world-class operations, or activity type (functional). Internal benchmarks are used when a company examines the processes and outputs of other departments, divisions, or subsidiaries. This involves significant information sharing between groups and in the process tells what resources are available internally, attracts support from colleagues, and helps to focus external benchmarking efforts. The results of internal benchmarking often lead to immediate gains. Competitive benchmarks involve investigating a competitor's services and processes. It is generally used for positioning products or services in a marketplace. Typically the technologies are similar and any lessons learned can be usefully adapted or translated. World-class operations benchmarking extends to examining companies who are excellent at doing specific things of significance to your own organization. Many processes and systems are generic and have cross-over value from one industry to another. Activity type (or functional) benchmarking is directed at process steps or discreet activities. Activity benchmarking can take place across dissimilar industries.

When adapting benchmarking results, they should not be used blindly. The lessons learned should be adapted, translated, or adjusted to suit the specific processes or situations. The information gained should be seen as a catalyst for originality, inspiration, or creativity.[73]

Also, it should be noted that benchmarks could be obtained from other than world-class performers. The mistakes and incidents of others could, for example, serve as warnings to be investigated and followed-up on.

Ross[74] describes various errors that can be made in the benchmarking process. These errors are:
- not involving the employees who will use the benchmarking information;
- not adequately defining your own process before gathering benchmark data;
- not seeing that benchmarking is not a one-time process, but is continuous;
- not expanding the scope of companies studied to include others outside your geographic or competitive area;
- not realizing that benchmarking is just a means to foster process improvement;
- not setting goals to bridge the gap to the desired performance;
- not empowering employees to achieve the improvements identified; and
- not quickly following-up on benchmarking to maintain momentum and keep credibility.

In a similar listing of benchmarking problems, DeToro[75] adds those of:
- lack of sponsorship;
- selecting the wrong people for the team;
- managers not understanding the necessary commitment;
- focusing on metrics rather than processes;
- not positioning benchmarking within a larger strategy;
- misunderstanding the organization's mission, goals, and objectives;
- assuming every project requires a site visit; and
- failing to inspect the benchmarking process.

C. BENCHMARKING PROCESSES

Various types of benchmarking models are found in the literature. For example, Spendolini[76] uses a model that has five steps. These steps are:
- determining what to benchmark,
- forming a benchmarking team,
- identifying benchmarking partners,
- collecting and analyzing benchmarking data, and
- taking-action.

In addition, Spendolini[77] reminds us that benchmarking is a continuous improvement process. The benchmarking model is a cycle repeated over and over.

Camp reports that others have described benchmarking processes in terms of from 4 to 12 steps.[78] His recommended model is:

- deciding what to benchmark,
- identifying who to benchmark,
- planning and conducting the investigation,
- determining current performance gaps,
- projecting future performance levels,
- communicating findings and gaining acceptance,
- revising performance goals,
- developing action plans,
- implementing the actions and monitoring progress,
- recalibrating the benchmarks.

These all take place in a process of planning, analyzing, integrating, and taking action.

D. SAFETY BENCHMARKING

Johnson and Burke[79] report on an example of safety benchmarking at the Sun Company where a study was begun to determine how its material safety data sheet (MSDS) management process compared to others. Consultants first identified industry leaders who were then sent detailed questionnaires and visited. The industry leaders were found to use high technology systems, including photographic scanning for their MSDS management programs. Final recommendations included: electronic data entry, use of advanced software, automatic generation of letters requesting MSDS from vendors, and a full regulatory database to keep up-to-date on regulatory changes.

Petersen[80] cites the results of benchmarking eighteen companies in the areas of safety leadership, safety performance, communication, and safety training. The best practices in safety leadership were:

- clear definitions of roles and responsibilities,
- having a champion for each important safety initiative,
- having visible top management ownership of safety, and
- constant identification and use of the company's best practices.

The best practices in safety performance were:

- using measures that drive continuous improvement such as observations,
- clearly defining and communicating performance targets,
- tying compensation and operating budgets to performance, and
- consistent application of disciplinary guidelines.

The best practices in communication were:

- soliciting feedback from the recipients of safety information and
- using a variety of media to deliver timely and consistent safety messages.

The best practices in training were:

- continually tracking training requirements,
- soliciting feedback from the course attendees,
- evaluating safety training at delivery and afterwards,
- using behavioral observations to determine training retention, and
- using innovative training techniques such as computer-based, self-paced training and incident re-creation.

THE BEST TECHNIQUES: ASSESSMENT QUESTIONS

- Has a work safety baseline been determined to identify the hazards and control methods used?
- Are world-class (best, better than required, etc.) work safety techniques from other organizations identified and applied to improve safety?
- Is a complete benchmarking process used to conduct the studies and implement actions?
- What efforts are made to identify best/better techniques from both inside and outside the organization for potential improvements?
- Have improvements been made in this way?
- Do all employees strive to help identify areas for improvement?

XIII. CONTINUOUS IMPROVEMENT

Continuous improvement means that there is continuous quality/safety improvement through the cycle of problem identification and analysis, development and implementation of corrective recommendations, review of results, and development of effective controls. The focus is on the factors which lead to accidents and injuries.

Improvements made should be both incremental and breakthrough. Improvements must be embedded by being part of the daily work of all units. They should be in processes seeking to eliminate problems at the source. And all improvements should be driven by opportunities to do better. Improvement can be:

- enhancing value to customers;
- reducing errors, defects, and waste;
- improving cycle time and responsiveness;
- improving productivity and effectiveness in use of resources; and
- improving the company's leadership position in fulfilling public responsibilities.

Improvement efforts tend to fail for one of five root causes. These are:

- Upper management not really believing that they need to change.
- A lack of trust between management and the employees. This is said to be the most frequent cause of failure, and has been reported to be a problem at up to 65% of companies examined.
- A poor choice of the improvement champion. The wrong person is picked for the wrong reasons.
- Improvement process based on a consultant's methodology where the wrong consultant or an incomplete method was selected.
- Forgotten middle-management. This is the group most suspicious of most improvement efforts.

Winchell[81] states that the most successful improvement efforts focus on the organization, the organization's processes and to a lesser extent, on individuals. Changes in these areas create permanent change in how the business is run. Example improvements could be changes in specific manufacturing activities to improve assembly, or changing the culture from production oriented to quality oriented, or in providing individual training for operators.

Organizationally, continuous improvement could mean a gradual change in how activities are conducted. Winchell[82] gives the example of an organization gradually shifting from

- a QC-applied inspection philosophy to one where QC and then the operators use SPC;
- then continuing to change to one where different functions work more closely together;

- then adopting training to foster top management and organizational cultural change;
- then adopting a team-based philosophy using process and project improvement teams leading to cross-functional work teams;
- and finally reaching a total customer-focus culture where all TQM principles and techniques are in place and functioning.

Process improvements, both in business and production processes, can be made more easily than organizational ones and are themselves long lasting. Typically, once the most important business processes are selected by a management steering committee, cross-functional performance improvement teams are established to determine what the difficulties are and how to improve them. Usually, cross-functional teams are required because most business processes involve many functions or groups, with many interfaces between them at which problems occur. An important part of the process improvement effort is in establishing process measurements which can be analyzed and trended when changes are made.

Individual improvement comes about through training, better job-matching, and the definition of what the customer-driven performance standards are. In addition, the individual needs to know what to do if the performance standards are not achieved.[83]

CONTINUOUS IMPROVEMENT: ASSESSMENT QUESTIONS

- Is there a record of continuous safety improvement through the cycle of problem identification and analysis, development and implementation of corrective recommendations, review of results, and development of effective controls?
- What kinds of safety improvements have been made?
- Are the results of improvement efforts tracked and reviewed?
- How often are changes made in the basic work processes?

XIV. QUALITY MANAGEMENT

Quality (or safety) management means using a worldwide standard quality management system for consistent quality and safety via effective policies and procedures.

A. QUALITY MANAGEMENT

There are many different quality management systems in use. Greene, for example, discusses 24 distinct approaches to Total Quality used worldwide.[84] Four effective quality management systems are: Socio-Technical Systems, Statistical Process Control, Just-In-Time, and ISO-9000.[85] In any company a quality management system may be organized by the particular standards used; the departments, functions, or organizations involved; or the business segment included.

- **Statistical Process Control:** These control systems establish process control limits and respond to variations exceeding those limits. The types of analysis and control tools used are described in Chapter 2, Section XI.
- **Socio-Technical Systems:** This system focuses on social (employee interpersonal) and technical systems and relationships. It aims to develop an open organizational structure that is able to learn and adapt. The process of establishing a socio-technical system involves:
  - examining each part of the management system as part of whole,
  - building a management program on high principles,
  - having all employees participate in work process improvement activities,

- ensuring that all work is designed as value adding, and
- striving so that the program is not just good enough but looks toward ideal solutions.
- **Just-In-Time Systems (JIT):** In this system the exact number of units is brought to each stage of production at the appropriate time. The program aims to produce to exact demand, continuously improve, and eliminate waste. Key components are improvement teams, vendor relations, and quality at the source.
  - **Improvement teams** are used to implement JIT. They are organized to improve the organization, build a better workplace, and develop individual knowledge, skills, and abilities.
  - **Vendor relations** has as a goal the establishment of long-term partnerships with single-source suppliers. The aim is to turn them into suppliers who provide the highest quality products while continually reducing costs. The results of vendor relations are reduced paperwork, prices negotiated based on cost analysis, and supplier checked and certified quality.
  - **Quality at the source** means that each operation views the next operation as a customer - endeavoring to do the job right the first time. Each worker is responsible for quality. There are no quality control inspections, there is respect for the worker's judgement, and if things go wrong, the process stops immediately to check for defective parts/equipment, overproduction (waste), or safety hazards.
- **ISO-9000 Systems** are simply management systems based on the ISO-9000 family of quality standards. The application of each of the 20 ISO-9001 quality management requirements to quality and safety management is thoroughly described in Chapter 3.

B. SAFETY MANAGEMENT

Ideally, any world-class quality management system could be used for safety management purposes. In this book the quality management systems used are TQM and ISO-9000. These represent a combination of the best quality management concepts (TQM) and the most widely accepted quality management framework (ISO-9000). In addition to these standards, the safety management standards promulgated by OSHA in the form of the VPP[86] and PSM[87] guidelines should be used to round out safety management requirements. The VPP and PSM guidelines represent proven standards of safety excellence implemented across the whole industrial spectrum.

QUALITY MANAGEMENT: ASSESSMENT QUESTIONS

- Has a comprehensive quality and safety management system including policies, processes, procedures, and standards been developed and installed?
- Are all the facets of the management system effectively implemented?
- How well can the managers and employees describe the work safety management system and how it is organized?

XV. QUALITY PLANNING

Quality (or safety) planning is planning to produce quality in lieu of prevention by inspection. To this end, quality planning tools allow detailed plans to be developed, communicated, and understood in sufficient detail for projects to be completed successfully. Planning tools and techniques include the following:[88]

(1) Affinity Diagram: This is group development of ideas and issues followed by their arrangement into affinity families. The ideas can be generated by brainstorming. Once all the ideas are arranged into groups, headers are created to describe the families. Affinity diagrams are good for large, complex issues with little structure.

(2) Relationship Diagram: This is basically an affinity diagram with arrows drawn to show the causal relationships between ideas. The diagram is group-generated to show all the items that relate to an issue or problem. The arrows show any causal relationships that exist. Outward groupings of arrows show the root causes of problems, while groupings of inward arrows show issues requiring investigation.

(3) Tree Diagram: This diagram depicts the logical sequencing of tasks needed to do a job. It begins with a goal or objective, and then proceeds by delineating the steps or requirements needed to achieve the goal. New levels of required steps result in filling in the diagram. This is similar to the Why-Why diagram discussed in section XI, B.

(4) Prioritization Matrix: Three types of matrices are used to prioritize tasks, options, and outcomes. Each type requires developing a weighting of options by criteria or comparing the options to each other.

(5) Matrix Diagram: This diagram identifies correlations of items from one group to another and shows strength of correlation. In this diagram, one set of tasks, ideas, or issues is listed vertically, and another set is listed horizontally. A team decides on the strength of the relationships and indicates that strength on the chart. This type of chart is often used to decide on responsibilities for tasks identified in a tree diagram.

(6) Process Decision Program Chart: This chart is used for detailed planning of every chain of events when doing something for the first time or when the goal is unfamiliar. It anticipates things that might go wrong. It starts with a tree diagram taken to one or two levels and proceeds by asking "What can go wrong?". For each answer, a plausible response is developed. The analysis can conclude with the development of a complete contingency plan.

(7) Activity Network Diagram: It is used to schedule the sequential implementation of a complex project. A sequential flow of tasks needed to complete activities is created, and then the paths are connected as appropriate. Times are allocated for each task and the earliest start and latest finish times are calculated. The diagram is used to monitor progress and the schedule. Typically:

- Tools 1 & 2 are used to generate ideas and define relationships.
- Tools 3, 4 & 5 are used to organize and prioritize.
- Tools 6 & 7 are used to plan for implementation and make decisions.

Other planning tools are the Gnatt Chart and the Storyboard.[89]

- **Gnatt Charts:** These are project scheduling tools that document the schedule, activities, and responsibilities needed to complete a project. The charts visually portray the sequence of events, when activities begin and end, and progress relative to that expected.
- **Storyboards:** These are used to provide an organized, visual summary of improvement projects. They highlight the logical steps in the project and follow by documenting and communicating results. As the project continues, findings and recommendations are displayed as well as events, flowcharts, and other summary data.

QUALITY PLANNING: ASSESSMENT QUESTIONS

- Is there an effective planning program to identify the critical and common cause factors of accidents and illnesses and to prevent accidents prior to occurrence for all of our work processes and activities?
- Are any specific analysis tools/techniques used in planning for work safety?
- Is staff trained to effectively use the quality planning tools?
- Is the responsibility for planning spread throughout the organization?

XVI. ASSESSMENT AND PLANNING

Assessment and planning are the initial stages of the improvement effort. They include activities that establish the present state of the organization in relation to its vision. Then they include developing a plan to achieve continuous improvement within an effective quality management structure. The major steps in the assessment and planning process are: conducting a thorough quality/safety assessment, completing a strategic plan with top managers heavily involved in setting a direction for action, and then establishing a tactical (or implementation) plan with objectives to achieve the organization's vision.

A. QUALITY/SAFETY ASSESSMENT

The quality/safety assessment identifies strengths, weaknesses, and areas needing improvement in the present programs; identifies opportunities for customer growth and satisfaction; identifies threats to success; and recommends goals for Critical Success Factors (CSF), projects, and training. CSFs are the factors that determine the company's probability of success in customer satisfaction, profit, growth, and competitiveness. They can be important measures of success, such as the percentage of on-time shipments or the lost-time injury rate, or they could be important activities needed to ensure satisfaction such as the servicing process.

The quality/safety assessment utilizes all relevant information to completely define the present status of the organization. For a safety assessment included would be:

- All company accident, injury, illness, and near-miss event data, analyses, and trends.
- General and industry-specific accident, injury and illness data.
- Previous safety management plans, programs, and results.
- The organization's safety measurement systems.
- Safety audits, reviews, and inspections, particularly recommendations for corrective and preventive action.
- Past and planned-for safety and safety management improvements.
- Employee grievances, suggestions, and satisfaction surveys.
- Interviews with top management to identify problems and improvement opportunities.
- Focus group meetings with middle and first-level management and employees to identify the present organizational culture, improvement needs, and obstacles.
- A comprehensive safety management audit/review.

B. STRATEGIC PLANNING

Strategic planning is planning to establish a direction and course for the company. The strategic plan defines the company's vision, mission, and values statements. As defined by Hradesky:[90]

- **Vision** is the state of being that the company desires to achieve in 3 to 5 years.
- **Mission** is a description of the products and services in relation to the customers and market and a description of the company's purpose, identification, and distinctive qualities.
- **Values** are observable beliefs, attitudes, and behaviors required to achieve the vision and mission.

Executives and top managers must participate in consensus and team building in this planning process.

Most companies using TQM have specific goals in mind. The recent Conference Board report[91] found that the three TQM objectives cited by more than 50% of the business units surveyed were: improving external customer satisfaction, reinforcing a culture of continuous improvement, and improving work processes. Other objectives cited by 20 to 40% of the respondents were: integrating quality with business practices, improving productivity/lowering costs, reducing cycle time to market, heightening employee empowerment, increasing teamwork, and improving long-term financial results. Similar types of goals would be central to any safety improvement effort.

50

Greene[92] cautions that TQM strategies need to recognize any cultural biases that U.S. organizations tend to exhibit. These biases favor launching a program over finishing the effort; communicating and thinking more than doing and implementing; relying on technical fixes rather than social changes; having extrinsic rather than intrinsic motivation; and using written rather than oral communication. Each type of bias can derail a TQM and likewise a safety change effort.

In planning to implement TQM, organizations should try to ensure that changes take place by replacement. This means replacing the old style of doing by the new quality style. As Greene[93] points out, implementation does not mean doing everything at once. Sequential changes avoid overusing resources and promote overall success by establishing a record of accomplishment.

C. TACTICAL PLANNING

Tactical or implementation planning creates an implementation road map (generally over two years) with objectives to achieve the goals. The tactical plan describes the CSFs, identifies necessary projects, sets out plans for the projects, and starts the team building process.[94]

Typical topics included in the planning could include the following:
- developing and fully communicating the vision, mission, and values statements;
- designing the CSF;
- implementing cultural change;
- choosing projects whose objectives impact on the CSF;
- developing high return-on-investment projects;
- developing projects required to close the gap toward the vision;
- establishing top level committees;
- establishing recognition and reward systems;
- designing training to complement project implementation;
- implementing tactical techniques such as SPC, ISO-9000, etc.;
- establishing a quality/safety management system, such as ISO-9000, complete with policies, procedures, and work statements.

Tactical planning should encompass all the elements needed to establish the new quality/safety system. Included in tactical plans are such elements as:
- business modeling;
- communicating the vision, mission, and values;
- developing departmental mission statements;
- establishing resource, executive, performance appraisal, and training committees;
- setting up the rewards and recognition system;
- establishing quality teams;
- and implementing quality management methods.

Planning should ensure that making the transition to the new quality/safety system is orderly and does not create performance and customer satisfaction problems.

In addition, tactical planning should recognize that making changes takes time and should be done right rather than quickly. Too often the tactical plan does not recognize organizational and personal constraints and requirements. The plan should be established recognizing:
- resource availability,
- ongoing activities,
- normal or expected workload fluctuations,
- union involvement,
- production requirements, and
- the time required to make changes to procedures and processes.

ASSESSMENT AND PLANNING: ASSESSMENT QUESTIONS

- Has our development and implementation of the work safety management system included a complete program assessment, strategic planning to develop new courses of action by executives, and tactical planning with safety projects and objectives?
- Have we developed safety vision, mission, and values statements and planned to implement a continuous improvement and effective process management structure?
- Have our vision, mission, and values statements been clearly communicated to staff?
- Have we planned to establish safety projects, committees, and teams to implement the safety vision and mission?
- Was the complete organization represented in the safety assessment and planning?

XVII. IMPLEMENTATION AND ORGANIZATION

Implementation and organization means that top management is fully committed to the quality/safety change process and has established the organization necessary to fulfill the strategic and tactical plans. The organization has a top-level quality or safety council with committees for resources, training, recognition and reward, and cultural change. Organization also means that there are established safety projects and project (safety) teams established and new processes, systems, and techniques implemented. All departments must be aligned with the organization's direction and should have internal customer satisfaction requirements, department mission statements, department requirements identified, department measurements of satisfaction, and full internal agreement on meeting expectations. The initial implementation tasks are to develop departmental mission statements and internal customer satisfaction agreements.

A. QUALITY/SAFETY COUNCIL

A strong and effective corporate quality/safety council is vital for the quality/safety change process to take place effectively. Based on the guidance given by Clemmer,[95] this council would provide input to and refine the implementation plan; develop the principles, policies, and guidelines for the improvement process; review and approve the roles of local quality/safety councils; become the champion and communicator of the improvement process; coordinate cross-functional initiatives; help to establish priorities; work with senior executives to resolve system and strategic process problems; track and measure the effectiveness of the improvement process; ensure that feedback loops are strong and effective; publish an annual report of progress; and develop and actively manage the reward and recognition system. This group should be chaired by a senior executive to signal top management commitment.

B. TEAM ORGANIZATION

The best organization is one where work safety is based on a decentralized, interactive system that integrates all levels with structural building blocks based on small teams, not big functions. Teams maximize involvement, provide agility and help to ensure an ownership focus. Teams are kept at manageable sizes, and each team is provided its own identity, with its own product. Interfaces between teams are clearly identified and each team is provided with ample authority over its own part of the product. In a team-based environment, safety professionals, providing functional safety expertise, must be part of the cross-functional work teams organized to improve safety.[96]

Petersen[97] lists steps for establishing effective safety teams.
- Management must define what decisions it will still make.

- Management and the worker must have a mature, professional attitude.
- A good level of trust must exist between management and workers.
- Ground rules must be defined for what is allowed and not allowed.
- Flexibility is permitted within the ground rules.
- Each team should have a periodic safety plan with a periodic effectiveness check.
- Each team is held accountable for its agreements.
- Determine the core elements of the safety plan but allow flexibility in implementation.
- Have intermediate self-checks and adjustments.

For the teams engaged in specific projects, a productivity/quality improvement process should be used consisting of:
- project identification, planning, and reporting;
- performance measurements;
- problem analysis and solution;
- an inspection/test capability study (can the process be adequately measured?);
- a process capability study (can the process be adequately controlled?);
- a process control procedure;
- a corrective/preventive action matrix;
- process control implementation;
- problem prevention;
- performance accountability; and
- measurement of effectiveness.

The output goals should be developed iteratively with the teams directly involved. The goals should be made understandable, relevant, attainable - and wanted. Provide ample incentive for initiative, ingenuity, and innovation.

In *Why TQM Fails*,[98] eight characteristics of high performance teams are listed. These are:
- clear, elevating goals;
- results-driven structure;
- competent team members;
- unified commitment;
- collaborative climate;
- standards of excellence;
- external support and recognition; and
- principled leadership.

IMPLEMENTATION AND ORGANIZATION: ASSESSMENT QUESTIONS

- Have we established an effective safety organization and work safety management system?
- Do we have:
  - safety projects?
  - project teams?
  - new processes, systems, and techniques?
- Are all departments aligned with:
  - internal requirements?
  - department mission statements?
  - measurements of satisfaction?
  - agreements on meeting expectations?
- Is there a strong and effective corporate safety council, chaired by a senior executive?
- Are work safety teams organized and used for involvement, agility, and an ownership focus?

XVIII. CULTURAL CHANGE

Cultural change means creating the commitment from the top to TQM (or safety) and transforming the company culture. During the cultural change process, values implementation is monitored and management empowers the workforce to improve processes and performance. The process of empowering employees encourages employees to take responsibility and ownership of problems and their solutions.

Instituting cultural change means:
- identifying the organization's values, both those to retain and those to adopt;
- defining the values;
- converting the values to behaviors;
- developing implementation plans for those values;
- executing the implementation plans;
- measuring the values implementation; and
- following-up to ensure achievement and sustainability of the values.

Creech[99] gives a list of activities required to firmly establish the character and culture of an organization. These activities are: "Develop the overarching principles. Key the principles to the human spirit. Ensure the principles are wholly understood and widely practiced - by all. Give them vigor through insistence, persistence and consistency. Stress ethical conduct, integrity, and courtesy in all endeavors. The principles flow from top down, but their power flows from bottom up."

In TQM (and safety) Hradesky[100] defines the most important cultural changes as:
- Shifting from bottom-line emphasis to "quality and safety first" equal to delivery and cost.
- Shifting from short-term objectives to a long-term view plus short-term objectives.
- Shifting from a "limits" satisfaction to continuous improvement by meeting stepped goals.
- Shifting from product delivery to continuous satisfaction, and a 0% fail rate.
- Shifting from quality-delegated responsibility to a management-led improvement process.
- Shifting from an inspection orientation to a prevention orientation.
- Shifting from sequential engineering to concurrent or parallel engineering.
- Shifting from management-by-edict to employee participation with the focus on common goals.

Management's responsibilities for cultural change are to assess and understand the existing culture, determine what culture is desired, develop implementation plans, become role models, provide expectations to all employees and empower them, create a culture-fostering group to monitor the progress of values, and provide adequate recognition and reward systems.

Management sustains cultural change by:
- maintaining a sense of urgency to sustain the desired changes;
- providing a clear-cut vision of the desired results;
- removing rewards that promote behavior opposite to that desired;
- showing the people what is wanted by example and message;
- having all of them do what is wanted;
- reinforcing desired behaviors by recognizing and rewarding them;
- expecting breakdowns and providing assistance to correct them;
- encouraging people to take risks and keep trying;
- rewarding success and training people in how to handle mistakes; and
- continually expressing commitment to the success of the people doing the work.

CULTURAL CHANGE: ASSESSMENT QUESTIONS

- Has there been a transformation of the work safety culture to one in which all basic TQM principles are embodied so that quality and safety are equal in importance to product delivery and cost?
- Is there is a management-led improvement process and do all employees participate with a focus on common goals?
- Is safety as important to the executives as are production, quality, and profit?
- Is safety as important to the employees as production and getting the work done on time?
- Has management instituted safety excellence awards?
- Has our work culture changed to one of safety by prevention rather than inspection?

XIX. RECOGNITION AND REWARD

Recognition and reward means that systems are established for recognizing and rewarding quality (and safety) and improvements in quality processes. The basis is that what gets rewarded and recognized gets repeated. Suggestions are rewarded. Performance appraisals are based on quality results and quality behavior implementation. Managers are accountable for quality. A blame-free atmosphere is established for errors. In TQM, the recognition and reward system communicates sincerity and commitment - demonstrating that everyone shares in the risks and rewards. Harrington[101] describes recognition and reward as a process that provides the reinforcement of desired behaviors to all managers and employees.

Appropriate recognition and reward address and reinforce quality and improvement results and behaviors and include quality-oriented performance appraisals focusing on job performance, process improvements, behavior implementation, and role model effectiveness. For example, Geller[102] suggests rewards should be based on both safety behaviors and accomplishments. Creek[103] notes that rewards should only be based on performance toward positive goals. He believes that rewarding workers for having accidents (even if fewer) is intrinsically wrong.

The keys to effective reward and recognition are cited by Clemmer[104] as follows:
- closely align rewards and recognition with the data-based management system;
- give everybody lots of information and feedback;
- involve those rewarded in deciding whom else should be rewarded and for what;
- ensure that senior executives are highly visible in the recognition process;
- make recognition as immediate as possible;
- make sure the compensation system is perceived to be fair; and
- don't allow recognition to focus only on the big successes.

Brown et al.[105] describe some common TQM measurement mistakes. These include reward systems which reward for behaviors rather than accomplishments, for courtesy rather than competence, and for making internal improvements while neglecting external improvements.

A. PERFORMANCE APPRAISALS

Specific recommendations for designing TQM performance appraisals are given by both Brown et al. [106] and Prince.[107] According to Brown, performance appraisals should be based both on team and individual goals. Effective performance appraisals should be open-ended and team-based using open-ended questions which merge planning and feedback during team meetings. The appraisal meetings should be designed around questions on past performance, future goals, developmental needs, and support needs. The timing of the appraisals should be aligned with the work (quarters, milestones, etc.). Appraisals should eliminate individual ratings, provide individual and team feedback, and use a natural way to document results.

Prince recommends that the rating system should be simple, with perhaps only three evaluation categories; that 360-degree appraisals, using input from self, peers, subordinates, superiors, and customers should be used; that the focus should be on work planning and development; that there be frequent performance review discussions; and that there should be a distinction between current job performance appraisals and any consideration for promotions - which should be based on future responsibilities.

The value of performance appraisals was verified by a survey of recognition and reward strategies conducted by The Conference Board.[108] The survey found that up to 79% of the companies and business units responding cited performance appraisals as the most consistently effective method of advancing TQM efforts for professional level personnel. Performance appraisals were also used at the executive level (75%), the non-exempt level (60%), and the hourly worker level (57%). It was also found that the business units with the best results incorporated about 100% more quality measures into their performance appraisal systems at all levels. This result was consistent with the use of other types of rewards.

There were also differences found in the types of individual performance measures used by the best practices units compared to the others. At the executive/management level, the best business units used more of the following types of measures: behavioral competencies, business processes, cycle time, defects/errors, employee involvement in quality, external involvement, and team performance. At the professional level, more of the business units with the best practices used defects/errors and quality leadership measures than did comparable units.

B. RECOGNITION

Recognition is an inexpensive, easy to use motivational technique which should be actively employed to foster improvement. Forms of recognition are team recognition and celebration, personal thanks, or individual recognition programs. In a team-based environment, recognition should be by peer group nomination. Recognition can be spurred by providing forums for employees to appear before peers to present improvement ideas and receive achievement awards.

Intangible rewards have great power and usefulness. Nelson[109] cites five "I" methods of no-cost, effective job recognition. These methods are:

- Interesting work;
- Information, communication, and feedback;
- Involvement in decisions;
- Independence with autonomy and flexibility; and
- Increased visibility and opportunity.

Petersen[110] cites ten research studies which note that recognition has been shown to be a very effective motivational tool for safety. The studies indicate that significant gains in performance were achieved by giving praise and feedback for using safe behaviors, reducing hazards and injury rates, and improving housekeeping.

C. INCENTIVES

Incentive reward programs are traditionally used in safety management in a effort to spur improved safety performance. However, if these programs are not well thought out, the programs can have detrimental effects such as employees not reporting injuries and illnesses when the result might be that a potential incentive or prize could be lost. This unintended effect damages not only the safety program and the work safety culture, but it also results in safety problems not being investigated and corrected, potentially leading to more serious incidents later.

In a recent article, Geller[111] reviews the state-of-the-art on incentive programs. He reports that contrary to the beliefs of many safety practitioners, safety incentive programs work if correctly constructed and implemented. He lists seven guidelines necessary for effective incentive programs.

These guidelines are that:

- rewardable and achievable safety behaviors should be specified;
- everyone meeting the reward criteria should be rewarded;
- small rewards for everyone are better than large rewards for a few;
- rewards should be displayable and represent safety achievement;
- rewards to one group should not be at another's expense;
- groups should not be penalized for an individual's failure; and
- progress should be monitored and posted.

These guidelines were successfully implemented in an incentive program that gave employees credits for exhibiting specific safety behaviors, such as attending safety meetings, leading safety meetings, conducting environmental audits, writing job safety analyses, and special participation.

D. COMPENSATION NOTES

The traditional focus on individual goals in performance appraisals can be destructive to quality and teamwork. It pits individuals against each other, forces ratings into a bell curve, reinforces the upright organizational pyramid, constrains sources of feedback, and increases inflexibility.[112] Thus performance-based compensation systems, if not constructed appropriately, tend to fail because:

- employees become more competitive not team oriented, or
- team members may all get the same bonus although only some did the work, or
- individuals and teams may exceed their goals but not get a bonus because of poor company results, or
- employees perceive the bonus incentive as not great enough to warrant increased effort, or
- employees figure out loopholes in the system.[113]

Compensation systems should strongly link quality (work safety) performance with pay, when appropriate. Gainsharing, the awarding of bonuses to employees based on the ability of teams to improve aspects of their performance can work well in a TQM environment.

RECOGNITION AND REWARD: ASSESSMENT QUESTIONS

- Is a recognition and rewards process established for employee safety and for improvements in the safety of the workplace and processes?
- Is everybody in the organization (managers, supervisors, and employees) clearly held accountable for meeting their safety responsibilities?
- To what extent do performance appraisals focus on job performance?
- Do our compensation systems have any linkage with work safety performance?
- Is my input sought on performance appraisals?
- Are our safety rewards team-based rather than individually-based?
- Are safety rewards equivalent to our quality and production rewards?

XX. LEADERSHIP DEVELOPMENT

Leadership development means that all employees are educated for leadership in quality (or safety) and that the organization is developing effective leaders to support TQM implementation. In the organization, the right jobs are performed at the right time, and employees and managers share in leadership and in objectives. Leadership training should be spread throughout the organization.

As noted by Hiam,[114] IBM gives leadership training to managers and employees to foster two-way communication while allowing each group to participate in the cultural change process.

TQM leaders can be and are developed through directed teaching and empowerment. This means that prospective leaders are given formal classroom training and a variety of diverse and increasing leadership opportunities. Most formal development programs make extensive use of simulations, scenarios, and modeling to make the educational process realistic and germane. Role models and mentors are also important to the leadership development process.[115]

Leadership is crucial to quality and safety. It supplies direction, vision, mission, and values, and it continually monitors progress and makes adjustments when needed. Strong leadership and commitment bring success. Leaders share visions, motivate employees, ensure that individuals are assessed based on performance, reward for positive behaviors, and select the trainers. Quality and safety leaders are the heros and champions who see quality (or safety) as a means to accomplish the company's long-term goals.

Many writers have examined and attempted to describe the characteristics and habits of leaders. Some examples of their syntheses are presented below.

A. LEADERSHIP CHARACTERISTICS

The characteristics of effective quality leaders are described by Hradesky.[116] Quality leaders show:

- Vision - the ability to see an end and/or goal in complete form;
- Confidence - knowing that you and your team can accomplish assigned tasks;
- Risk taking - the willingness to try new methods;
- Decision making - making the right, tough, and courageous decision;
- Development of others - the ability to create leadership thinking and action in others;
- Influence on others - imparting others with energy and enthusiasm;
- Communication - the ability to channel ideas into action.

As described by Covey,[117] leaders (effective people) exhibit seven common habits.

- Leaders are proactive - they take the initiative and responsibility to make things happen.
- Leaders begin with an end in mind - they know and work toward the end point.
- Leaders put first things first - they organize and execute around priorities.
- Leaders think win-win - they obtain agreements and solutions that benefit and satisfy all parties.
- Leaders seek first to understand and empathize - they build trust through careful listening.
- Leaders synergize - they create solutions through the application of all the leadership habits.
- Leaders renew - they enhance themselves and their natures, physically, emotionally, mentally and spiritually.

Creech[118] contrasts leadership and managership. In his view, leaders are the proactive drivers and shapers of the organization, while managers are the more passive pilots and controllers. In specific:

- " Leaders shape the outputs. Managers chase the inputs.
- Leaders focus on group projects. Managers focus on individual jobs.
- Leaders encourage new ideas. Managers enforce old ideas.
- Leaders stimulate the right things. Managers monitor for the wrong things.
- Leaders thrive on tough competition. Managers talk little of competition.
- Leaders prize comparison with others. Managers see scant need for comparison.
- Leaders think of improvement programs. Managers think of suggestion programs.
- Leaders empower others to make decisions. Mangers tightly control the decision process.
- Leaders see leading as animate and proactive. Managers see managing as inanimate and reactive.

- Leaders think of a dynamic, caring, human system. Managers think of a business, following a script.
- Leaders think of improving initiative and innovation. Managers think of improving compliance and conformance.
- Leaders shape the organization's character, culture and climate. Managers assume that's neither a big deal - nor their job."

Or in very concise terms,

- " Leaders create the vision; managers carry out the vision.
- Leaders make it better; managers make it run.
- Leaders make it happen; managers hope it happens.
- Leaders create more leaders; managers create more managers."

B. SAFETY LEADERSHIP

All of the leadership characteristics and habits outlined above apply directly to safety leadership. For example, when safety is considered, the seven leadership characteristics described by Hradesky translate into:

- Vision - the leadership vision is one of no accidents in a harmonious and safe workplace.
- Confidence - that workers can meet safety objectives while completing projects and jobs.
- Risk taking - the willingness to question the status-quo and try new methods to foster continuing safety improvement.
- Decision making - making the hard safety decisions such as those for corrective discipline or halting production for a safety concern.
- Development of others - developing other safety leaders by example, coaching, and delegation of responsibility and authority.
- Influence on others - giving others the desire to work safely and help improve the system.
- Communication - continually communicating the safety vision and culture, while coaching subordinates and peers.

In a direct application of leadership characteristics to safety, Lark[119] describes safety leadership in terms of the acronym FIRM where:

- **F** stands for faith in seniors, subordinates, and self;
- **I** stands for integrity and idealism;
- **R** stands for respect for others and for self; and
- **M** stands for manners and moderation.

FIRM safety professionals believe in the job, set a good example, coach others, maintain self-discipline, and say "thank-you."

FIRM non-supervisors have the ability to identify concerns, report the concerns to others, suggest solutions, consult with others, and even follow-up on the corrective action. They have faith in themselves, want the problem solved, and trust others to help.

FIRM supervisors have faith in safety - exemplified by their behaviors; have integrity - continuously demonstrating commitment; respect the employees' ability to work safely and make contributions; and have the manners to reward and praise safety excellence.

FIRM executives (as described by Lark) af-FIRM and con-FIRM their commitment to safety by exhibiting positive leadership and creating the organizational climate for safety success.

LEADERSHIP DEVELOPMENT: ASSESSMENT QUESTIONS

- Have we developed and educated all of our employees about safety leadership to help implement the work safety management system?
- Do we develop leaders who share visions, motivate employees, and ensure that individuals are assessed based on performance?
- Do we develop leaders who focus on results, build employee commitment, effectively lead meetings, and create positive working relationships?
- Is safety leadership only left to the safety group?
- Do employees, identify concerns, report concerns to others, suggest solutions, consult with others, and follow up on corrective actions?
- Do supervisors continually demonstrate safety behavior, show a commitment to safety, and reward and praise safety excellence?
- Do our leaders have a safety vision, have confidence in the staff, take risks to improve safety, make hard safety decisions, develop other safety leaders, influence others to work safely, and continually communicate the safety message?

XXI. TEAM BUILDING

Team building means that quality/safety teams are developed, established, and educated in effective team work concepts. Teams are integral to work safety improvement because they provide the leadership necessary to achieve company goals. For effective teams, selecting the right members and leaders from the beginning is critically important. The right combination of members brings a balance in technical and business skills and develops synergies. Well-run teams can accomplish a lot more, quickly, and generally at less cost than a single individual.

Hradesky[120] defines the four main stages in team building as occurring in the following order: forming, storming, norming, and performing.

- Forming is the process of becoming aware and oriented to the team setting. It includes goal setting, commitment, acceptance of the limits to action, and rules of behavior.
- Storming is the stage of conflict and resistance to team goals and tasks. In storming, people criticize, boast, and fight for control.
- Norming is when communication and cooperation begin to develop. This stage includes respecting others, listening and learning, and generally supporting the team's efforts.
- Performing is the process of producing and solving problems. At this final stage, team members work together to achieve set goals for a feeling of accomplishment and pride.

A. TEAM COMMUNICATION

Strong two-way communication is the key to developing strong teams. To develop strong communication ideas should not be judged - they should be evaluated; no team member should feel superior to others - each member brings certain strengths to the team; no member should try to control others - each member should be free to express his ideas; there should be no manipulation of team members; team members should not feel indifferent toward the team - they should be supportive of team objectives; and team members should not feel certainty about ideas and efforts - they should be open-minded about the ideas of others.

Team communication is achieved by:

- describing ideas;
- treating others as equals;

- being open to others;
- staying problem oriented;
- keeping a positive attitude;
- being understanding of others;
- seeing situations from others' point of view; and
- being trusting.

B. TEAM SKILLS

Team members need to adopt behaviors that optimize the team's effectiveness. Members have to learn to use improvement tools and techniques and learn how to contribute to the team; leaders must learn how to lead and build teams. The important skills for team members and leaders are:
- data-based tools and techniques;
- process management;
- problem solving; and team member and participation skills.

Team leadership skills are:
- facilitating successful meetings (focusing on quality and getting a team committed);
- making sure the meetings mirror the culture; and
- advanced coaching skills.[121]

C. TEAM PITFALLS

Pitfalls in teams are:[122]
- not preparing the group to be a team;
- not selecting a strong leader;
- not holding team members accountable;
- selecting a process no one is interested in improving;
- not having team members who will benefit from the solution;
- not having team members who can technically contribute;
- not having regular weekly meetings (unless there is a valid reason);
- selecting the desired solution instead of studying the process;
- not publishing weekly progress reports;
- not putting measurements in place;
- not having process management conducted by higher levels of management.

Other pitfalls are:[123]
- not fostering and renewing team skills in top managers;
- not ensuring good team skills in teams;
- reverting to command and control under pressure; and
- having a weak top management team.

TEAM BUILDING: ASSESSMENT QUESTIONS

- Have we established safety (and project) teams?
- Are all team members educated in effective team work concepts, including the importance of two-way communications?
- Have we taught the team members the skills needed to successfully function on a team?
- Have we carefully selected strong team leaders?
- Are teams and team members clearly accountable for their performance on the team?
- Do the teams successfully work on safety projects?
- Does each safety team have input or a member from the safety function?

XXII. HIRING AND PROMOTING

Hiring and promoting means that all hiring and promoting decisions are based on acceptance of the quality/safety culture and on quality/safety performance factors equally with others.

A. HIRING

Hiring represents the ideal opportunity for an organization to make sure that it obtains personnel who will fit in with the quality/safety culture that it is promoting. Thus it is incumbent for managers to take all the time necessary and use all the resources necessary to hire those that will succeed. To paraphrase Congreve's quotation,[124] Hire in haste, we may repent in leisure. Only in this case, the leisurely repenting might involve serious organizational harm. With this in mind, interviews should be planned and conducted to obtain as good an assessment as possible of the applicant's quality/safety beliefs and values, as well as his technical knowledge and proficiency. The interview process should be thorough and tough. Managers should only hire those who have the basic qualities they want. In addition, Krause[125] suggests that valid screening tests should be used to evaluate factors such as dependability, personal relations, and safe work habits.

The immediate manager should be personally and closely involved with the selection process, not delegate it to the personnel department. To the extent possible, other managers should participate in the interviews to get alternative viewpoints and to build organizational trust and foster teamwork. References and past employers should be checked. Sincerity in the TQM effort is demonstrated by the management selection process. Since managers train, mentor, reward, and enforce - sincerity in their selection is a powerful message of organizational commitment to the workforce.

Hiring success does not end with the making of an offer and having the new employee arrive for work. To ensure a smooth initiation to the organization, there must be an effective new employee orientation program. Effective orientation results in fewer mistakes, improved performance and service, higher levels of productivity, and more harmonious employee relations. A good orientation also communicates a message to the new employee. It tells him that the organization cares for him and is serious about its quality/safety culture.

B. PROMOTING

Both promoting and hiring send a vital message about the quality/safety culture to the entire organization. Promoting someone who violates quality/safety rules, immediately tells everyone that the quality/safety is really not as important as some other factors. No matter what management says or does otherwise, promoting a rules violator will seriously harm a positive quality/safety culture.

Promoting managers is a critical element in the quality/safety effort. Since the managers will train, mentor, reward and enforce in keeping with the quality/safety culture - sincerity in their selection will send a powerful message of leadership commitment to the workforce and help ensure that the desired organizational values are continued.

All promotions, demotions, and dismissals should be made in complete alignment with the quality/safety culture. Clemmer[126] lists the five key factors used for promotion in a business environment. Adapting them to quality/safety gives:
- quality/safety results;
- quality/safety leadership;
- human resources management;
- teamwork; and
- adherence to the organization's quality/safety values.

These same factors should be an integral part of the performance appraisal process, so that promotion decisions are based on a solid history of factual information.

HIRING AND PROMOTING: ASSESSMENT QUESTIONS

- Do we make a major effort to hire only those who will fit in with the organization's quality/safety culture?
- Are our hiring interviews tough and demanding?
- Are the line managers personally and strongly involved in hiring?
- Does the orientation program stress the safety culture, safety programs, and safety expectations?
- Are all of our promotion decisions based on acceptance of our safety culture and safety performance, equally with other criteria?
- Are all promotions in alignment with safety results, leadership through safety and quality, human resources management, teamwork, and corporate safety values?
- Does the safety function have any input on promotion and hiring decisions?

XXIII. MANAGEMENT READINESS

Management readiness means that managers must be developed to become more efficient and effective in achieving the desired business results and in fostering the crucial quality or safety management behaviors and objectives. First line supervisors must not only perform the usual management functions, they must be trained in the skills essential for the new safety culture.

A. MANAGEMENT SKILLS
Hradesky[127] defines the crucial skills needed by managers in a quality/safety environment as:
- Observing and interpreting behavior for hiring and coaching personnel.
- Coaching for results and behavior.
- Managing multiple priorities (time management).
- Running results oriented quality/safety meetings.
- Conducting quality/safety oriented performance appraisals.

In a different way, Castillo[128] describes six skills as important to the safety manger in improving performance. In his viewpoint, the important skills are:
- Kaleidoscopic thinking. The ability to rearrange and reassess data in new ways. Creative thinking to apply knowledge beyond traditional boundaries.
- Communicating ideas and visions.
- Persistence. Time is a key factor in projects that succeed. Projects must be pushed beyond roadblocks.
- Coalition building. Provides support for ideas to receive acceptance and resources.
- Working through teams. Change must be implemented by more than one person.
- Sharing credit.

B. COACHING
Coaching is fundamental to safety/quality improvement. In safety oriented organizations, supervisors and peers continually coach individuals to correct mistakes and use appropriate safety behaviors. The four functions of effective coaching are listed by Hradesky[129] as:
- Counseling (description of problems, technical and organizational insight, changes in points of view, commitment to self sufficiency, and deeper personal insight).
- Mentoring (political savvy, organizational culture, networking, self management, commitment to organizational goals, and sensitivity to senior managers).

- Tutoring (increased technical competence and understanding, movement to expert status, increased learning pace, and commitment to continued learning).
- Confronting (clarification of expectations, identification of performance deficits, acceptance of more difficult tasks, strategy to improve performance, and commitment to continued self-improvement).

Coaching skills needed are:
- giving recognition, encouragement, and positive feedback;
- frequent quality feedback on work safety performance;
- work safety performance management with high uncompromising standards;
- fostering continuous work safety improvement and innovation;
- managing change;
- focusing individuals and teams on safety and quality;
- using core coaching skills; and
- training and developing team members.[130]

Geller[131] uses the work "coach" to describe the essential coaching skills as follows:
- **C**-communications,
- **O**-observation,
- **A**-analysis,
- **C**-change, and
- **H**-help.

Common pitfalls to developing coaching skills are:
- confusing inspiration and education with skills,
- assuming that managers have coaching skills just due to their experience,
- underdevelopment via one-day workshops,
- cutting out front-line supervisors/coaches from service/quality improvement strategies,
- not aligning the whole management system with coaching,
- failing to coach the coaches, and
- executives not leading by example.[132]

MANAGEMENT READINESS: ASSESSMENT QUESTIONS

- Are all our managers (especially first line supervisors) trained in the skills essential for the safety culture?
  These skills are:
  - observing and interpreting safety behaviors,
  - counseling (or coaching) for safety,
  - running results-oriented safety meetings,
  - conducting safety-oriented performance appraisals.
- Are all our supervisors ready to perform the functions of:
  - analyzing the work they supervise to identify unrecognized potential hazards,
  - maintaining physical protection in their work areas,
  - reinforcing employee training on the nature of potential hazards and needed protective measures?
- Are the supervisors ready and able to coach for improved safety through communicating, observing, analyzing, changing, and helping?
- Do the managers and supervisors effectively apply these skills?

XXIV. TOTAL QUALITY TRAINING

Total quality training is necessary to provide the bases, elements, and supportive knowledge for all quality/safety characteristics and concepts. Training must take place in all aspects of quality/safety. Vital also is the development of effective trainers and training programs for continued implementation and improvement. Education ensures that the people in the team-based, continually improving, customer-focused organization will see, understand, and embrace those ideals. Training consists of developing an understanding of what is to be done and why (education) and the ability to make targeted changes happen (skills). For continuous quality improvement, the learning process must also be continuous and long-term.

Creech[133] summarizes the key points of training for quality. "Provide detailed, focused training to employees at every level. On-the-job and ad hoc training are key parts, but are only parts. Formal training is vital for proper quality mindset and know-how. Make all training specific on key principles, methods and goals. Train all employees at every level - including senior levels. Leaders at all levels must be teachers. Leaders create leaders."

These points are further emphasized in the Conference Board Report, *Does Quality Work?*[134] This report describes a study by Katheryn Troy in which 13 leading TQM practitioners agreed that effective TQM training requires:
- targeting training on those who can use the training to lead and train others;
- involving all layers of managers and supervisors in training;
- stressing timely on-the-job application of training;
- tailoring curricula and materials to the specific business;
- using employees as trainers; and
- using alternative training technologies.

Introducing quality/safety begins with the managers and the process should not proceed until they are all educated and aware and have developed the coaching and team leadership skills required.

Personal Skills: The general success of the improvement effort will be in direct proportion to the investment in skill development. The skills needed are technical, data-based tools and techniques to improve the processes, and interpersonal, human, or people skills. This last category is vital, for 80% of quality/safety problems are related to front-line interpersonal, management, leadership, support, and involvement skills. The personal skills needed are:
- working effectively;
- communicating clearly, and helping others do the same;
- providing positive mutual support;
- seeking and accepting help when needed; and
- constructively channeling emotion.

Brown et al.[135] outline what it takes to make training successful. For a successful program, it is recommended that the training cover:
- Quality/safety concepts; using relevant examples (actual situations), tailoring training to the organization, and following up with concrete actions.
- Quality tools; providing a framework to use skills in the work environment.
- Special topics; training employees in the right areas, organizing courses into a logical curriculum, developing suggested training paths for each major group of employees, establishing curriculum design early.
- Leadership training; providing ongoing feedback.

In terms of these ideas, Brown et al. also describe why training fails.[136]
- Quality concept; employee skepticism, unrealistic expectations about just what training will

accomplish, training not tailored to audience, training thought to be irrelevant, employees doubt company commitment, training does not start at the top, and lack of follow up.

- Quality tools; not enough practice (need lecture, demonstration and practice, sometimes repeated practice for complex tools), and lack of applicability or understanding of how the tools apply to the job.
- Special topics; systematic needs analyses not conducted and courses not logically aligned.
- Quality leadership; using antiquated theories, using pop psychology models, relying on self-discovery tests, and lack of applicability and follow up.

TOTAL QUALITY TRAINING: ASSESSMENT QUESTIONS

- Does our training program provide everyone with the bases and elements for all work safety characteristics and concepts, and for continual implementation and improvement in work safety programs?
- Does our safety training program include the following elements:
  - work safety concepts?
  - safety/quality tools and techniques?
  - leadership?
- Does training include both technical and personal skills?
- Are managers, supervisors, and employees used as trainers?
- Is training targeted to the job and applied to the job?

XXV. SUMMARY - TQM SAFETY SYSTEM EVALUATION

Total Quality Management Topic	Rating*
1. Product and Customer Focus	
2. Leadership Commitment	
3. Company Culture	
4. Effective Communication	
5. Organizational and Employee Knowledge	
6. Employee Empowerment	
7. Employee Responsibility and Excellence	
8. Management by Fact	
9. Long-range Viewpoint	
10. Statistical Process Control	
11. Structural Problem Solving	
12. Best Techniques	
13. Continuous Improvement	
14. Safety Management	
15. Safety Planning	
16. Assessment and Planning	
17. Implementation and Organization	
18. Cultural Change	
19. Recognition and Reward Systems	
20. Leadership Development	
21. Team Building	
22. Hiring and Promoting	
23. Management Readiness	
24. Total Quality Training	
Total Safety Program Score	

TQM SAFETY SYSTEM RATING GUIDE

(Rate each topic 0 to 4. Consider the scope and breadth of deployment and the results achieved.)

> *4= Basically all of our people or activities meet the stated criteria or their intent with excellent results achieved.*
> *3= Most of our people/activities meet the criteria or their intent with good results.*
> *2= About half of our people/activities meet the criteria or their intent with positive results.*
> *1= Some or a few of our people/activities meet the criteria or their intent with a few positive results.*
> *0= None of our people/activities meet the criteria and no positive results are evident.*

TQM SAFETY SYSTEM EVALUATION GUIDE

Total Score   *Safety System Evaluation*

81-96   *Our company fosters world-class work safety characteristics. TQM characteristics are very evident in safety activities with very few or no gaps or problems.*

65-80   *Our company has very progressively built TQM into safety with sound approaches and few gaps in deployment and integration, but some fine tuning is possible in several areas.*

49-64   *Our company is progressive in building TQM into safety but gaps in deployment exist and refinements are still needed.*

33-48   *Our company has made progress in applying TQM to safety but significant gaps in deployment exist. Improvement is still needed throughout.*

17-32   *Our company is in the early stages of building TQM into safety and requires substantial improvement in many areas.*

0-16   *Our company is very traditional and reactive toward safety management with limited success potential. The application of TQM principles to safety may be just beginning.*

XXVI. REFERENCES

1. *Malcolm Baldrige National Quality Award 1996 Award Criteria*, American Society for Quality Control, Milwaukee, WI, 1995, 30.
2. **Smith, T. A.**, Managing for continuous improvement in safety performance - focus on the customer, *Professional Safety*, 40.2, 18, 1995.
3. **Ferry, T.**, *Safety and Health Management Planning*, Van Nostrand Reinhold, New York, NY, 1990, Ch. 15.
4. **Petersen, D.**, *Analyzing Safety System Effectiveness*, Van Nostrand Reinhold, New York, NY, 1996, 27.
5. **Creech, B.**, *The Five Pillars of TQM*, Truman Talley Books/Dutton, New York, NY, 1994, Chap. 5.
6. **Nelson, A. J.**, Remarkable zero-injury safety performance, *Professional Safety*, 41.1, 22, 1996.
7. *Malcolm Baldrige National Quality 1996 Award Criteria*, 30.
8. **Clemmer, J.**, *Firing on all Cylinders - The Service/Quality System for High-Powered Corporate Performance*, Business One Irwin, Homewood, IL, 1992, 101-107.

9.   **Clemmer**, *Firing on all Cylinders*, 340.
10. **Clemmer**, *Firing on all Cylinders*, 341.
11. **Harrington, H. J.**, *Total Improvement Management*, McGraw-Hill, New York, NY, 1995, 65.
12. **Krause, T. R., Hidley, J. H., and Hodson, S. J.**, *The Behavior-Based Safety Process: Managing Involvement for an Injury-Free Culture*, Van Nostrand Reinhold, New York, NY, 1990, 64-67.
13. **Nelson**, Remarkable zero-injury safety performance, 25.
14. **Harrington**, *Total Improvement Management*, 68.
15. **Greene, R. T.**, *Global Quality: A Synthesis of the World's Best Quality Methods*, Irwin Professional Publishing, Burr Ridge, IL, 1993, 133-134.
16. **Simon, R.**, The trust factor in safety performance, *Professional Safety*, 41.10, 28, 1996.
17. **Hradesky, J. L.**, *Total Quality Management Handbook*, McGraw-Hill, New York, NY, 1995, 12.
18. **Kelly, F. R.**, Worker psychology and safety attitudes, *Professional Safety*, 41.7, 14, 1996.
19. **Roughton, J.**, Integrating a total quality management system in safety and health programs, *Professional Safety*, 38.6, 32, 1993.
20. **Petersen**, *Analyzing Safety System Effectiveness*, 68.
21. **Petrick, J. A, and Scherer, R. F.**, *Organization Ethics Development, Health and Safety*, The Minerva Education Institute, Education Module CM95-01, 1995. (Available at http://www.minerva.com.)
22. **Creech**, *The Five Pillars of TQM*, 325.
23. **Petersen**, *Analyzing Safety System Effectiveness*, 116-117.
24. **Creech**, *The Five Pillars of TQM*, 530.
25. **Larkin, T. J. and Larkin, S.**, Everything We Do Is Wrong, in *The Quality Yearbook - 1996*, Cortada, J. W. and Woods, J. A., Eds., McGraw-Hill, New York, NY, 1996, 396.
26. **Geller, E. S.**, 20 guidelines for giving feedback, *Industrial Safety and Hygiene News*, July 1996., Internet: www.safetyonline.net/ishn/9607/behav.htm.
27. **Roughton**, Integrating a total quality management system, 35.
28. **Brown, M. G., Hitchcock, D. E., and Willard, M. L.**, *Why TQM Fails And What To Do About It*, Irwin Professional Publishing, Burr Ridge, IL, 1994, 215.
29. **Brown**, *Why TQM Fails*, 216-227.
30. **Greene**, *Global Quality*, 129.
31. **Greene**, *Global Quality*, 194-195.
32. **Solomon, C. M.**, The Learning Concept: How It's Being Implemented, in *The Quality Yearbook - 1996*, 385.
33. **Biech, E.**, *TQM for Training*, McGraw-Hill, New York, NY, 1995, 106.
34. **Bell, C. R. and Zempke, R.**, *Managing Knock Your Socks Off Service*, AMACOM, New York, NY, 1992, 157.
35. **Hradesky**, *Total Quality Management Handbook*, 161.
36. **Hradesky**, *Total Quality Management Handbook*, 162-163.
37. **Biech**, *TQM for Training*, 106-109.
38. **Hayes, B. E.**, How to measure empowerment, *Quality Progress*, 27.2, 41, 1994.
39. **Topf, M. D. and Petrino, R. A.**, Change in attitude fosters responsibility for safety, *Professional Safety*, 40.12, 24, 1995.
40. **Harrington**, *Total Improvement Management*, 272.
41. **Harrington**, *Total Improvement Management*, 308-311.
42. *Malcolm Baldrige National Quality Award 1996 Award Criteria*, 31.
43. **Creech**, *The Five Pillars of TQM*, 479.
44. **Hradesky**, *Total Quality Management Handbook*, 65-66.
45. **Clemmer**, *Firing on all Cylinders*, 275-277.
46. **Brown**, *Why TQM Fails*, 84-85.
47. **Brown**, *Why TQM Fails*, 58-62.
48. **Pfau, B. N. and Gross, S. E.**, *Innovative Reward and Recognition Strategies in TQM*, Report No. 1051, The Conference Board, New York, NY, 1993, 16.
49. **Clemmer**, *Firing on all Cylinders*, 261-271.
50. **Brown**, *Why TQM Fails*, 66.
51. **Roughton**, Integrating a total quality management system, 36-37.
52. **Krause**, *The behavior-based safety process*, 41-43.

53. **Krause**, *The behavior-based safety process*, 45-46.
54. **Krause, T. R.**, *Employee-Driven Systems for Safe Behavior*, Van Nostrand Reinhold, New York, NY, 1995, 78-79.
55. **Geller, E. S.**, Safety coaching: key to achieving a total safety culture, *Professional Safety,* 40.7, 16, 1995.
56. **Grimaldi, J. V. and Simonds, R. H.**, *Safety Management*, 5th Edition, Irwin, Burr Ridge, IL, 1989.
57. **Johnson, D. and Burke, A.**, Turn up the heat, *Industrial Safety and Hygiene News*, October 1996, Internet: www.ishn.com/cov1096.htm.
58. *Malcolm Baldrige National Quality 1996 Award Criteria*, 31.
59. **Narus, B.**, *The Leader's Edge*, Contemporary Books, Chicago, IL, 1989, 124-127.
60. **Narus**, *The Leader's Edge*, 128-129.
61. **Capezio**, *Taking The Mystery Out Of TQM*, 250.
62. **Shearer, C.**, *Practical Continuous Improvement for Professional Services*, ASQC Quality Press, Milwaukee, WI, 1994, Chapter 7.
63. **Krause**, *Employee-Driven Systems for Safe Behavior*, Chapter 7.
64. **Krause**, *The Behavior-Based Safety Process*, 79-81.
65. **Creek, A. N.**, Organizational behavior and safety management, *Professional Safety*, 40.10, 36, 1995.
66. **Shearer**, *Practical Continuous Improvement*, 194-196.
67. **Shearer**, *Practical Continuous Improvement*, 157.
68. **Beich**, *TQM for Training*, 194-196.
69. Safety and Health Management Guidelines, U. S. Occupational Safety and Health Administration, Federal Register, 59:3904-3916, 1989.
70. Voluntary Protection Program: Part 1: Program Elements, DOE/EH-0433, U. S. Department of Energy, Washington, D. C., 1994.
71. **Spendolini, M. J.**, *The Benchmarking Book*, AMACOM, New York, NY, 1992, 9.
72. **Greene**, *Global Quality*, 120-121.
73. **Shearer**, *Practical Continuous Improvement*, 157-159.
74. **Ross, J. E.**, *Total Quality Management: Text, Cases and Readings*, 2nd Edition, St. Lucie Press, Delray Beach, FL, 1995, 249.
75. **DeToro, I.**, The 10 Pitfalls of Benchmarking, in *The Quality Yearbook - 1996*, Cortada, J. W. and Woods, J. A., Eds., McGraw Hill, New York, NY, 1996, 439-442.
76. **Spendolini**, *The Benchmarking Book*, 46-49.
77. **Spendolini**, *The Benchmarking Book*, 50.
78. **Camp, R. C.**, *Business Process Benchmarking*, ASQC Quality Press, Milwaukee, WI, 1995.
79. **Johnson**, Turn up the heat.
80. **Petersen**, *Analyzing Safety System Effectiveness*, 29-31.
81. **Winchell, W.**, *Continuous Quality Improvement: A Manufacturing Professional's Guide*, The Society of Manufacturing Engineers, Dearborn, MI, 1991, 53-54.
82. **Winchell**, *Continuous Quality Improvement*, 55-57.
83. **Winchell**, *Continuous Quality Improvement*, 61-62.
84. **Greene**, *Global Quality*, Ch. 3.
85. **Capezio**, *Taking The Mystery Out Of TQM*, Chap. 14.
86. Safety and Health Management Guidelines, U. S. Occupational Safety and Health Administration, Federal Register, 59:3904-3916, 1989.
87. Process Safety Management, OSHA-3132, U. S. Occupational Safety and Health Administration, Washington, D. C., 1994 (Reprinted).
88. **Beich**, *TQM for Training*, 191-193.
89. **Swanson, R. C.**, *The Quality Improvement Handbook*, St. Lucie Press, Delray Beach, FL, 1995, 179-191.
90. **Hradesky**, *Total Quality Management Handbook*, 12.
91. **Pfau**, *Innovative Reward and Recognition Strategies*, 12.
92. **Greene**, *Global Quality*, 110-113.
93. **Greene**, *Global Quality*, 117.
94. **Hradesky**, *Total Quality Management Handbook*, 18.
95. **Clemmer**, *Firing on all cylinders*, 297-298.

96. **Leisheid, W. E.**, TQM & safety: new buzz words or real understanding, *Professional Safety*, 39.6, 31, 1994.
97. **Petersen**, *Analyzing Safety System Effectiveness*, 156-157.
98. **Brown**, *Why TQM Fails*, 116.
99. **Creech**, *The Five Pillars of TQM*, 528.
100. **Hradesky**, *Total Quality Management Handbook*, 145.
101. **Harrington**, *Total Improvement Management*, 469.
102. **Geller**, Ten principles for achieving a total safety culture, *Professional Safety*, 39.9, 18, 1994.
103. **Creek**, Organizational behavior, 37.
104. **Clemmer**, *Firing on all cylinders*, 234-5.
105. **Brown**, *Why TQM Fails*, 58-62.
106. **Brown**, *Why TQM Fails*, 112-117.
107. **Prince, J. B.**, Building performance appraisal systems consistent with TQM practices, in *Human Resources Management Perspectives on TQM*, Knouse, S. B., Ed., ASQC Quality Press, Milwaukee, WI, 1996, 50-53.
108. **Pfau**, *Innovative Reward and Recognition Strategies*, 15.
109. **Nelson, B.**, Secrets of successful employee recognition, *Quality Digest*, August, 1996, Internet: w3.qualitydigest.com/~qdigest/nelson.html/#anchor212128.
110. **Petersen**, *Analyzing Safety System Effectiveness*, 75-76.
111. **Geller, E. S.**, The truth about safety incentives, *Professional Safety*, 41.10, 34, 1996.
112. **Brown**, *Why TQM Fails*, 109-110.
113. **Brown**, *Why TQM Fails*, 122-123.
114. **Hiam, A.**, *Closing the Quality Gap: Lessons from America's Leading Companies*, Prentice Hall, Englewood Cliffs, NJ, 1992, 227.
115. **Gibson, J. L., Ivanaevich, J. M., and Donnelly, J. H. Jr.**, *Organizations: Behavior, Structure, Processes*, Irwin, Burr Ridge, IL, 1994, 438.
116. **Hradesky**, *Total Quality Management Handbook*, 196-198.
117. **Covey, S.**, *The Seven Habits of Highly Effective People*, Simon and Schuster, New York, NY, 1989.
118. **Creech**, *The Five Pillars of TQM*, 303-304.
119. **Lark**, Leadership in safety, 34-35.
120. **Hradesky**, *Total Quality Management Handbook*, 215.
121. **Clemmer**, *Firing on all cylinders*, 194-197.
122. **Hradesky**, *Total Quality Management Handbook*, 223.
123. **Clemmer**, *Firing on all cylinders*, 206-207.
124. **Congreve, W.**, from The Old Bachelor, Act V., *Bartlett's Quotations*, www.cc.columbia.edu.
125. **Krause**, *Employee-Driven Systems for Safe Behavior*, 144.
126. **Clemmer**, *Firing on all cylinders*, 153-154.
127. **Hradesky**, *Total Quality Management Handbook*, 232.
128. **Castillo, J. E.**, Safety management: the winds of change, *Professional Safety*, 40. 2, 34, 1995.
129. **Hradesky**, *Total Quality Management Handbook*, 241-242.
130. **Clemmer**, *Firing on all cylinders*, 183-189.
131. **Geller**, Safety coaching, 17-21.
132. **Clemmer**, *Firing on all cylinders*, 191-192.
133. **Creech**, *The Five Pillars of TQM*, 530.
134. **Hiam, A.**, *Does Quality Work? A Review of Relevant Studies*, Report 1043, The Conference Board, New York, NY, 1993, 18.
135. **Brown**, *Why TQM Fails*, 40-41, 49-53.
136. **Brown**, *Why TQM Fails*, 42-49.

Chapter 3

IS0-9000 REQUIREMENTS

In this chapter the requirements of the ISO-9000 family of standards are described and adapted to the management of occupational or work safety programs. The primary ISO documents used are ISO-9001[1] and ISO-9004.[2]

ISO-9000 topics described are:

- Management Responsibility
- Safety Management System
- Contract Review
- Design Control
- Document and Data Control
- Purchasing
- Control of Customer-Supplied Product
- Product Identification and Traceability
- Process Control
- Inspection and Testing
- Control of Inspection, Measurement, and Test Equipment
- Inspection and Test Status
- Control of Nonconforming Product
- Corrective and Preventive Action
- Handling, Storage, Packaging, Preservation, and Delivery
- Control of Quality Records
- Internal Quality Audits
- Quality Training
- Servicing
- Statistical Techniques

Note: Most of the background and discussion material is based on ISO requirements and recommendations in ISO-9001 and ISO-9004-1. Although this material is discussed in terms of a quality management system and requirements, the same ideas apply to a safety management system. Where it is appropriate in the discussion, specific safety and safety management applications are described. All of the assessment questions are phrased in terms of safety management.

I. MANAGEMENT RESPONSIBILITY

Management has the responsibility to ensure that a quality system exists and is effective. In quality terms, management responsibility means that management is responsible for:

- defining the policies and objectives for quality;
- specifying the responsibilities and authority of all personnel whose work affects quality;
- appointing a person responsible for the quality system;
- ensuring the quality system is effective; and
- demonstrating a commitment to quality.

These same requirements apply identically to establishing and implementing a safety management system.

A. QUALITY POLICY

The written quality policy is the lynchpin of all organizational quality efforts. The policy expresses, usually in a clear and direct statement, the organization's quality aspirations and commitments. Once expressed, the policy is then made concrete by the development and implementation of quality objectives and a quality management system.

There are important steps which should be followed by management in establishing a quality (or a safety) management system. These steps are as follows. Management should:

- assign responsibility for someone/some team to develop the quality policy;
- ask for input from across the company to ensure "ownership" of the quality goals;
- develop comprehensive objectives;
- consider organizational goals; and
- in the policy, include a commitment to quality, products, services, customers (expectations and needs), safety, continuous improvement, and responsibility to society.

The quality policy should: be written; contain objectives for quality and commitment to quality; be relevant to organizational goals and customer needs; and be understood, implemented, and maintained throughout the organization. The policy should also be consistent with other company policies, written and unwritten.

It is vital that all personnel understand the quality policy and can express what it means in their own words. This is achieved by constant reinforcement through communication. Management should develop a plan to ensure that the policy is understood, implemented, and maintained at all organizational levels. The policy should be covered in new employee orientation, copies should be displayed, and it should be discussed in departmental meetings. Management should reinforce and follow up on the ideas in the policy, and verify its effectiveness.

B. QUALITY OBJECTIVES

Objectives restate the quality policy in more specific terms that show how important quality really is. They should be established for all departments and groups. In practice, these top-level objectives are implemented through even more specific goals for departments and groups.

For safety, program objectives recommended by the National Safety Council[3] are a starting point for consideration. These objectives include:

- gaining and maintaining program support at all levels of the organization;
- motivating, educating, and training personnel to recognize, correct, and report hazards;
- engineering hazard control into the design of machines, tools, and facilities;
- having an inspection and maintenance program for machinery, equipment, tools, and facilities;
- incorporating hazard control into training and educational techniques and methods; and
- complying with established safety and health standards.

For a continuously improving safety program, however, safety objectives should be more far-reaching. For example: safety program requirements should exceed those covered by OSHA regulations and applicable consensus standards; maintenance should strive for maximum reliability of equipment; and accident rates should be minimized and constantly lowered.

C. QUALITY RESPONSIBILITY

In establishing an organization, management should clearly define all responsibilities, authorities, and how the assignments are interrelated. Management should prepare organizational charts and should review and expand job descriptions for personnel whose work affects quality and who have authority over identifying problems, generating solutions, initiating corrective action, verifying implementation of the corrective action, and controlling nonconforming product.

Note that for quality (and for safety), the clearly defined responsibilities and authorities include everyone in the entire organization. Emphasis should be placed on problem identification and timely corrective and preventive action.

Ferry[4] gives examples of safety responsibilities for various organizational levels and functions in a typical company. In his summary he states that:

- Senior executives set policy direction, review control information, delegate safety and health responsibility and authority, and make budgetary allocations.
- Line management conducts operator training, supervises workers, communicates on hazards, and consults with employees.
- Safety staff informs management of safety conditions, advises purchasing of new safety standards, and advises the personnel department on available safety training.
- Employees conduct work according to established rules, report unsafe conditions to supervisors, and report any injuries or accidents to supervisors.

In a similar fashion, Ferry also outlines the safety responsibilities of functional groups such as purchasing, design, finance, production, quality control, administration, maintenance, and sales/marketing.

D. QUALITY RESOURCES

Management should identify the resources necessary for management, work performance, and verification (audit and inspection) activities. These necessary resources include the personnel resources and specialized skills needed. Thus, management needs to define the level of competence, experience, and training necessary to meet quality program objectives. Other resources include those for design and development equipment; manufacturing equipment; inspection, test, and measurement equipment; and instrumentation and computer software.

E. MANAGEMENT REPRESENTATIVE

Management should appoint a management representative, generally high-level, with responsibility for ensuring the quality system is developed, implemented, and maintained. The management representative is also responsible for reporting to management on the performance of the quality management system in fostering improvement, and the management representative also liaises with outside organizations regarding the quality system.

F. MANAGEMENT REVIEWS

There should be periodic, comprehensive management reviews of the quality system, including the assessment of quality audit results, quality system effectiveness, the identification of defects and problems and their correction, and the maintenance of records. The reviews should also consider the need to update the quality system based on new technologies and standards, new concepts, or other conditions. Management should ensure that any necessary corrective actions stemming from the management review are supported and implemented.

MANAGEMENT RESPONSIBILITY: ASSESSMENT QUESTIONS

- Has management:
  - defined the policies and objectives for the safety management system?
  - specified the safety responsibilities and authority of all personnel?
  - appointed a person responsible for the safety management system?
  - ensured the safety system is effective?
  - demonstrated its commitment to safety management?
- Is the safety policy written in clear, unambiguous language?
- Is the safety policy consistent with other company policies?
- Does the safety policy refer to employee expectations?
- Can the employees explain the safety policy in their own terms?
- Is the safety policy made concrete by proactive safety objectives stressing continuous improvement and safety excellence beyond basic compliance?
- Does everyone have an approved job description that includes responsibility and authority for safety?
- Does the safety manager or executive have the clear authority and responsibility to implement the safety management system?
- Do performance appraisals include safety performance considerations?
- Are adequate resources provided for safety?
- Is there a periodic, comprehensive management review of the safety system and its effectiveness?
- Is the review conducted by independent individuals?
- Does the review include extra-regulatory issues?

II. SAFETY MANAGEMENT SYSTEM

In quality terms this means that a system that ensures that work is performed to specified requirements is established, documented, implemented, and maintained. The quality system is the sum of the organizational structure, responsibilities, procedures, processes, and resources for implementing quality management. A safety management system ensures that all work is performed safely, that desired levels of safety are maintained, and that safety program objectives are met.

Establishing a comprehensive quality system starts with determining the requirements of the system based on regulations and standards, including those for documentation and implementation. The structure of the quality system documentation needs to be planned including any manuals, operating procedures, instructions, records, forms, and specifications. After that, existing company practices should be determined by using and reviewing flowcharts, procedures (written and unwritten), and instructions. Present and future resources need to be evaluated, including those for personnel, instrumentation, specifications and acceptance standards, and quality records.

After evaluating resources, a plan should be developed to implement the quality system. Included in the planning process is identifying and obtaining the controls, processes, equipment, resources, and skills necessary; ensuring that all designs, processes, procedures, and documentation are compatible; updating control, testing, and inspection techniques; identifying any extreme measurement requirements; clarifying acceptance standards; and identifying quality records. Control elements in the quality system plan should include program reports, action item lists, audit reports, and findings, etc.

Johnson[5] describes the goals of a quality system. The quality system: must be well understood and effective; must be effective in meeting quality objectives; must provide confidence that the product meets customer expectations; must give emphasis to preventive action; and must be documented. Another goal is that no nonconforming product reaches the customer. These same goals apply to a safety management system.

The quality system should apply to all quality-related activities. Likewise the safety management system should cover all phases of safety-related activity. For quality, these phases and activities include: marketing and research; design specification engineering and product development; procurement; process planning and development; production; inspection, testing, and examination; packing and storing; sales and distribution; installation and operation; technical assistance and maintenance; servicing; and disposal after use.

An important aspect of the quality or safety system is that it includes plans for special or new projects or activities. These plans should ensure that all company quality or safety objectives are considered and include specific allocation of responsibilities and authorities; specific procedures and work instructions; suitable testing, inspection, examination, and auditing; a method for making changes and modifications; and any other measures needed to ensure quality or safety.

In augmenting the guidance in ISO-9001, ISO-9004 includes material on developing operational procedures, and documenting the quality system through written policies and procedures, typically using a quality manual. The quality manual is important as it brings all quality-related information together to document the system and serve as a reference. In practice, the form of the documentation used can vary to suit the size and structure of the organization.

Procedures in the quality manual should address all of the requirements in ISO-9001. Russell[6] cautions that procedures and instructions should be clear, correct, and effective. They should agree with the actual work flow and process, they should use correct and appropriate terminology, and they should contain criteria for success. Procedures should also be verified to be effective prior to use.

OSHA recommends[7] specific safety program elements. These elements include management leadership and employee participation, workplace analysis, accident and record analysis, hazard prevention and control, emergency response, and safety and health training. See Chapters 4 and 5 of this book for the two specific sets of safety management program standards promulgated by OSHA. These are the Voluntary Protection Program[8] and Process Safety Management.[9]

SAFETY MANAGEMENT SYSTEM: ASSESSMENT QUESTIONS

- Do we have a safety management system that ensures that work is performed safely: established, documented, implemented, and maintained?
- Have regulatory and other safety requirements (standards) been determined including: measurements, documentation, and implementation?
- Has the structure of the safety documentation been planned, including: manuals, operating and safety procedures, work instructions, records, forms, and specifications?
- Do the safety manual and procedures accurately reflect the safety program and practices?
- Are procedures clear about work to be done, safety hazards and precautions, safety requirements, and any necessary approvals?
- Have resources (present and needed) been evaluated, including those for personnel, instrumentation, specifications and acceptance standards, and safety management records?
- Does the safety management system include:
  - hazard prevention and control?
  - facility and equipment maintenance?
  - planning and preparing for emergencies?
  - a medical program?

III. CONTRACT REVIEW

Contract review means that contracts or orders from external customers are reviewed to ensure that the requirements are adequately defined and that the company can meet them. There should be no surprises and all customer expectations should be clear. Any agreed sale of a product or service constitutes a contract. In terms of safety, contract review has implications both for formal contracts to supply products and services and also for the implied contract with the workers.

Note that in the broad sense, the term contracts refers to purchase orders, sales agreements, requests for proposal or quotation, product or service descriptions in catalogs, advertisements, sales presentations, marketing brochures, data sheets, price lists, etc.[10]

A. SAFETY IMPLICATIONS

In terms of safety, the intent of the contract review section in ISO-9000 is broadened. In ISO-9000, it is enough that contract requirements are defined and understood, and that the company can meet the requirements of the contract. For safety, however, there are multiple aspects to consider. First, there is the direct application to **formal contracts** on products or services. For these, the company should review the contracts to understand what the safety implications of meeting them are and then ensure that all safety concerns and requirements are satisfied. These safety concerns could be external (affecting the purchaser and public) or internal (affecting the employees). Although much of this process takes place in the normal course of business, it is important that the safety implications of any changes to requirements are adequately understood and reviewed.

Secondly, there is the **implied contract** or agreement with the real customers of the safety management system - the employees. In this view, the expectations of the workers must be clearly defined and understood by the organization, and the organization should strive to ensure that these expectations are met. Worker expectations can be determined through focus groups, meetings, questionnaires, or other similar means. Also, a feedback process should be developed to ensure that problems or concerns are corrected so that the safety system really meets worker expectations.

Thirdly, Kozak and Krafcisin[11] extend the meaning of contract review to include the inputs that drive the safety program. In this sense, contract review would include laws and regulations, internal requirements, voluntary standards, labor union contracts, insurance company recommendations, and risk assessment and audit recommendations.

B. ISO IMPLEMENTATION

Fully implementing the requirements of an ISO-type safety management system requires that the developer understands: what the requirements of ISO-9000 are; what the present program and work practices are; what it takes to implement an ISO quality system; and how to mesh safety management ideas and requirements into the ISO-type program. Generally, present programs and practices are best determined and understood through a detailed analysis process which includes flowcharting all activities. Implementation might include developing and reviewing changed procedures, ensuring that proper controls are in place so that requirements are met, and obtaining agreements from all parties to the process both external and internal. All of these activities must take place to implement an effective contract review system.

C. CONTRACT REVIEW ELEMENTS

In ISO-9001, the major elements of contract review are: defining and documenting the customer's requirements before accepting a contract; ensuring that any verbal orders are clear and agreed to; verifying the capability to meet contract requirements; identifying how a contract amendment is made and how the amendments are communicated throughout the organization; and maintaining records of all contract reviews. The different types of customer requirements usually include performance characteristics, sensory characteristics, installation configuration, applicable standards and regulations, packaging, and quality assurance and verification.

CONTRACT REVIEW: ASSESSMENT QUESTIONS

- Are contracts or orders from external customers reviewed by a procedure that ensures that any requirements that may adversely impact on worker safety are identified and accounted for in the work processes?
- Are our contract reviews effective for standard and non-standard contracts, including verbal, orders?
- Do we document the customer's requirements including those for testing, shipping, labeling, documentation, and field and startup testing?
- Do we verify our capability to meet requirements?
- Are contract changes adequately considered and communicated throughout the organization?
- Regarding the implied contract with the workers, are the expectations of the workers defined and understood by the organization?
- Is there feedback from the workers on the safety system?
- Are safety program changes made which reflect worker feedback and concerns?
- Are all regulatory requirements, internal standards, and other guidelines factored into the safety management system?
- Are there current copies of regulations and standards available?

IV. DESIGN CONTROL

Design control means that product, process, and service design is planned, organized, and controlled so that the resulting design meets agreed quality (safety) requirements and needs.

A. DESIGN CONTROL ELEMENTS
The separate elements of design control include:
- design and development planned for all phases of design control, including definition of responsibilities for all activities;
- organizational and technical interfaces defined and information transmitted correctly, with documentation and proper review;
- design inputs identified, documented, and reviewed, with any conflicts resolved;
- design outputs documented with requirements for verification and validation, and identifying critical characteristics for safe and proper functioning;
- design review at appropriate design stages, with representatives of all design stage functions;
- design verification at stages to assure output meets input requirements;
- design validation to ensure user needs are met; and
- design changes and modifications identified, documented, and reviewed.

B. PLANNING FOR DESIGN CONTROL
Design and development plans should be realistic. All design and development activities should be done by qualified personnel with appropriate and adequate resources. Design planning should identify responsibilities for all design phases from initiation through design reviews and validations, including how design changes are documented. Plans should be revised as necessary to keep them current. Detail can be added as the work progresses.

C. DESIGN REVIEW

Design reviews, verifications, and validations differ. Reviews are formal, comprehensive, and systematic to evaluate design requirements and the capability to meet requirements. Personnel performing design reviews should represent all functions involved with that stage of the design. Design review elements include product specifications and service requirements, process specifications, and customer needs.

- Product specifications are such things as: reliability, permissible tolerances, acceptance criteria, ease of assembly, and labeling and warnings.
- Service requirements include maintainability and the ability to diagnose problems.
- Process specifications are such things as: manufacturability, testability, materials specifications, and packaging and handling requirements.

D. DESIGN VERIFICATION AND VALIDATION

Verifications are reviews, audits, inspections, tests, or checks to verify conformance to requirements. Verification is usually provided through the following types of activities: alternative calculations, comparison with proven design, qualification tests, trials and demonstrations, comparison of design characteristics crucial to safe functioning, review of design documents, facilities and software, and review of documents before release. In addition, analytical calculations are generally used when it is impractical to verify by testing.

Validations are tests to ensure products meet user needs or requirements. In contrast to verification tests, validation tests may require simulating actual use conditions which may differ from contract requirements.

E. SAFETY IMPLICATIONS

Designing for safety is critically important because once a production system is designed and installed, little improvement can be accomplished. Superior quality and safety results are attainable only if quality and safety are designed in or if processes (systems) are redesigned to achieve excellence. Safety hazards and quality defects are most effectively and efficiently anticipated, avoided, mitigated, or controlled during the initial stages of systems design or redesign.[12]

For safety it is important to include design control provisions that ensure:

- reviewing all customer agreements and applicable statutory and regulatory requirements at the beginning;
- reviewing the requirements to anticipate, avoid, mitigate, and control hazards and assign responsibilities;
- obtaining input from all cross-functional activities to establish interfaces, and review for safety and manufacturability;
- designing the output to meet input requirements, contain reference data, meet regulations, and consider safety needs. Making sure that statutory, regulatory, safety, and customer specified requirements are met.
- the design review is documented including participants representing all concerned functions and safety personnel at appropriate design stages;
- the design review includes product instructions, brochures, legal documents, and quality control that includes safety characteristics;
- engineering methods such as failure mode and effects analyses, fault tree analyses and human reliability analyses used as appropriate in the design review process;
- making sure that all necessary procedures are written clearly and are available where they are used;
- engineering controls and warning labels are included to point out hazards, environmental considerations, and safe working methods;
- product design includes labeling, wording for advertisements and cataloging, user

instructions, and servicing instructions; and

- all documentation is reviewed before release.

Ferry states that 70 to 90% of design decisions affecting safety are made at the conceptual phase of the design process.[13] During this phase, the designer needs to be aware of the product's intended use, how many will be produced, what the critical environments are, what the critical manufacturing steps are, and what the failure modes and consequences are. The designer should seek to avoid errors of omission and should be cognizant of all human aspects associated with the product such as abuse, assembly, maintenance installation, and operation.

Human factors engineering should be considered when reviewing designs. Of particular note should be identifying which functions must be performed by humans, what types of controls are needed to exercise human control, how controls are easily identified, and the availability of appropriate critical systems redundancy. When correctly applied, human factors engineering improves system effectiveness, leads to fewer performance errors and accidents, minimizes retrofit and redesign, reduces training cost, and leads to more effective use of personnel.

Petersen notes that the primary controls on safety during design are functions of the qualifications and performance of the design staff.[14] All design personnel need to be appropriately trained, all design and design review functions need to be effectively performed, and all supervisory and managerial responsibilities need to be carried out. Training of the design staff should include safety, industrial hygiene, ergonomics, and systems safety.

DESIGN CONTROL: ASSESSMENT QUESTIONS

- Is product, service and process design planned, organized, and controlled so designs meet safety management objectives for excellence?
- Is the design staff adequately trained and supervised?
- Do design control procedures cover:
  - plans for each design activity?
  - interfaces?
  - design input?
  - design output?
  - design review?
  - design changes?
  - verification?
  - validation?
- Have all customer agreements (input) been documented?
- Has design control been established to anticipate, avoid, mitigate and control hazards, and assign responsibilities?
- Do procedures to control and verify the design of products, processes, and devices ensure that statutory, regulatory, safety, and customer specified requirements are reviewed and met?
- Are these procedures available at all points of use?
- Are all the design control procedures effectively carried out?
- Are design safety reviews carried out by cross-functional teams including health and safety professionals?

V. DOCUMENT AND DATA CONTROL

Document and data control means that controls are employed to ensure that valid, up-to-date documents and data are used and that invalid documents and data are not used in processes and operations that affect quality and safety. In fact, all documents that can affect quality and safety should be suitably controlled.

Arnold states that many companies do not adequately describe what documents are to be considered as quality documents.[15] Inadequacies are often found in listing and controlling inspection procedures, test instructions, and operating procedures. This section of the ISO standard is generally the largest source of noncompliances.[16]

ISO requires: establishing and maintaining data and document control procedures; reviewing documents by authorized personnel prior to release; establishing a master list of all current documents; establishing a control plan for each category of document; and reviewing all changes to documents and data by the same functions that performed the original review and approval unless otherwise specified. Document control should reflect the type of document. As noted by Arnold,[17] there are two types of documents: those that are retained and not changed - such as test reports, and those that are changed - such as drawings and procedures. Documents that change should be clearly marked with the latest revision number.

A. MASTER DOCUMENT LIST

The documents on the master list include all forms, drawings, manuals (quality and safety), specifications, inspection procedures and instructions, test procedures, maintenance procedures, work instructions, and quality control and safety procedures with the current revision status and dates of the items on the list. Document control should also include test data, qualification reports, audit reports, instrument calibration data, and quality (safety) cost reports. Pertinent subcontractor documents such as installation instructions and maintenance requirements should also be controlled.

B. CHANGE REVIEW

Changes to documents should be made only after careful consideration of the documented reasons for change. The review should be done by the same functions and at the same level that originally reviewed and approved the document to ensure that there is a proper understanding of the document's purpose. Thus all background and reference material related to the change should be available to those doing the review. If allowed by the design control process, minor and limited scope changes to documents may be reviewed by personnel at lower levels. In this instance, the allowed extent of any document revision must be predetermined.

There should also be a formal process for making and approving urgent, temporary changes to controlled documents, effective for a limited time, quantity, or contract. Since the degree of control over such changes is limited, only authorized personnel should be allowed to make and approve temporary changes.

C. DOCUMENT AVAILABILITY, OBSOLESCENCE, AND REISSUE

It is important to assure the availability of documents at the operating locations or wherever they are needed. For example, local copies of documents should always be available and not kept in a locked office after normal working hours. Also, it is important to establish control over documents that become obsolete or out-of-date with prompt removal from places of issue or use, or appropriate marking. Document control should assure that any marked up documents are marked such as "uncontrolled copy - not to be used." If obsolete copies are retained, special controls should be used.

ISO-9004 also recommends that documents should be reissued after a set number of revisions. Since document revision processes usually involve cover sheets and discussions of reasons for changes, reissuing documents has the effect of making them easier to understand and use.

DOCUMENT AND DATA CONTROL: ASSESSMENT QUESTIONS

- Are controls employed to ensure that valid, up-to-date documents and data are used in all processes that affect worker safety?
- Is there a master list for all forms, drawings, manuals, procedures, and work instructions that affect safety with the current revision status and dates of the items on the list?
- Do we plan for the control of each category of document, document the procedures, and verify the review and approval of all documents?
- Do we ensure document availability at operation areas and establish control for obsolete or out-of-date documents with prompt removal or appropriate marking?
- Have we established, implemented, and verified change control procedures?
- Are temporary changes made with adequate review?
- Are all temporary changes made permanent in a timely manner?
- Are there any informal, uncontrolled documents like job aids at the workplace?
- Are all federal and state safety and health regulations maintained up-to-date?
- Does everyone who needs safety documents have up-to-date copies?

VI. PURCHASING

Purchasing means that controls are employed to ensure that purchased products or services that can affect quality or safety meet specified quality or safety requirements. Thus both the purchaser and the supplier must understand what the specified requirements are.

For safety, purchasing can be viewed in three separate ways. First, purchasing applies to materials or components which will be used in producing a product or in processing. In this instance, which is similar to the ISO intent, the purchaser must ensure that the safety of the final product or the safety of the process will not be compromised. The quality of the component or material purchased must be clearly specified and verified. The purchasing function should ensure that all servicing instructions are provided, any training that is required is provided, any hazard information is provided, proper labels and warnings are provided, and proper containers for hazardous substances are provided.[18]

In a second way, purchasing can apply to contracting for safety services or for purchasing safety products. Where safety programs involve using and purchasing services and materials such as air sampling/analysis, personal protective equipment, and other safety equipment obtained through outside suppliers, the quality of these products must be ensured.[19]

In a third way, purchasing can apply to the process of contracting for personnel who will work on-site either as full-time employees or as spot contractors.

For purchasing, ISO requires establishing procedures to ensure that purchased products meet specified requirements; evaluating subcontractors; defining the amount of control over subcontractors; establishing records of acceptable subcontractors; ensuring that purchasing documents clearly describe the product ordered; and specifying and agreeing on requirements for verification such as verification at the subcontractor's site. ISO-9004 also speaks to having procedures for settling quality disputes and planning for receipt inspections and controls.

Purchasing documents should be reviewed and approved prior to release. Purchasing data should clearly and specifically describe the product/service purchased including: the type, class, style, grade, or other identification; the title or other positive identification and applicable issue of specifications, drawings, process requirements, inspection instructions, etc.; and the title, number, and issue of the quality system standard that applies. Specific quality requirements should be specified, such as 100% inspection, submission of selective test data with the purchased products, lot inspection by the purchaser, or implementation of a complete QA program by the contractor. As a safety example, purchasing specifications for new machinery could include ergonomics, manuals, safety mechanisms, test runs, machine guarding, and testing laboratory listing.

Contractor acceptability should be based on quality and safety criteria. For a product, quality criteria include product quality history, delivery dependability, and quality system capability. Acceptability may be determined by inspection of the contractor's site and operations, by evaluating samples, or by reviewing published data. Safety criteria would also include the safety history of contractor products. The control exercised over subcontractors should depend on the type of product and on the product's impact on safety.

When purchasing or obtaining the services of contracted personnel to work on-site, it is important to determine the contractor's safety history. It is also vital to ensure that the contracted personnel are qualified and trained appropriately and that they will follow all plant quality and safety procedures. There should be a specific purchase order with clearly described safety requirements.

Acceptability verification of purchased products can take place at the subcontractor's place of business - or should be specified in the purchase order. Verification does not necessarily mean inspection, it only means that the product is verified to meet contracted specifications.

ISO-9004 recommends that receiving inspections and controls should be suitably planned for. The level of inspection and place of inspection should be determined, as should the availability of properly calibrated test equipment and supplies and properly trained personnel. As necessary, records should be provided to continually monitor and evaluate contractor performance.

PURCHASING: ASSESSMENT QUESTIONS

- Are controls employed to ensure that purchased products or services that can affect worker safety meet specified safety requirements?
- Have we evaluated purchasing specifications including material data safety sheets (MSDSs)?
- Have we evaluated suppliers and contractors based on their performance histories?
- Do we have an approved supplier and contractor listing?
- Are our purchase orders very specific about the identification and classification of items and any requirements that can affect safety?
- Have we reviewed the development and approval process?
- Do we ensure that the specifications for new machinery include
    - ergonomics?
    - manuals?
    - safety mechanisms?
    - test runs?
    - machine guarding?
    - testing laboratory listing?

VII. CONTROL OF CUSTOMER-SUPPLIED PRODUCT

Control of customer-supplied product means that the condition and security of customer-supplied products is controlled and the customer notified if necessary. Customer-supplied product is product owned by the customer and provided to the supplier for use in meeting requirements. The objective is that deterioration of the customer-supplied product should not affect quality and safety. In ISO-9000 this is a special requirement that refers only to materials supplied by the customer for use in the product which is ultimately going to the same customer.

A. SAFETY CONSIDERATIONS

For safety this also is a requirement of very limited applicability which applies to two types of situations. First, it would cover situations where workers supplied their own tools or materials for use in a plant, or where prior training was relied on to fulfill a safety commitment. In these situations the acceptability of the tool would have to be assured as would the appropriateness and completeness of the training.

Second, it also covers situations when product oriented materials are supplied by the ultimate customer and these materials might affect worker safety. Customer-supplied products could be materials, services (like transportation), packaging materials, labels, connectors, electrical components, software modules, or subassemblies. Other customer-supplied products could include fixtures, tools, or test and measurement equipment.

B. BASIC REQUIREMENTS

The basic ISO requirements are that there are procedures for the control of verification, storage, and maintenance of customer-supplied product which will be incorporated into the product or service. Any lost, damaged, or unsuitable material should be reported back to the customer (supplier of the material).

To implement these provisions, the supplier must establish procedures for the verification, storage, and maintenance of customer-supplied products; ensure the suitability of customer-supplied and worker-supplied materials for use; and establish a system of recording and reporting products damaged or unsuitable for use.

Implementing a customer-supplied product control process could include: establishing receipt inspections, putting customer-supplied product in a separate storage area, clearly marking the product as customer-supplied, and ensuring that everyone is aware of which material was supplied by the customer.[20]

CONTROL OF CUSTOMER-SUPPLIED PRODUCT: ASSESSMENT QUESTIONS

- Are the condition and security of customer-supplied products controlled so that safety processes are not adversely affected?
- Do we have procedures for the verification, storage, and maintenance of customer-supplied products?
- Do procedures establish a system of recording and reporting customer-supplied products that are damaged or unsuitable for use?
- Do the products covered by the system include materials, fixtures, tools, and measurement equipment.
- Are worker-supplied tools assured acceptable for safe use?
- Have prior worker training and qualifications been verified before relying on them?

VIII. PRODUCT IDENTIFICATION AND TRACEABILITY

Product identification and traceability means that processes, products, and services are suitably (and if necessary, uniquely) identified through all the stages of production, delivery, and installation so that quality (safety) objectives can be met.

Identification is the ability to distinguish one product, part, procedure, or process from another. Identification of parts allows workers to ensure that the proper parts are used in an assembly. Identification of procedures ensures that the proper procedures have been used for an activity or that the proper steps have been taken in an analysis. Identification of test data and instrument calibrations ensures that important test data is being analyzed properly. In all cases, the type of identification employed should suit the specific requirements, importance and risk levels.

Traceability is the ability to track the history, application, or location of an item or activity by means of recorded identification. For products, traceability may begin with receipt of raw materials, may identify all intermediate stages of production, and may continue to the final end user. Three aspects of traceability can be defined. These aspects are: input traceability (sources of products and services), operational traceability (such as process characteristics, employee effectiveness, and equipment maintenance), and traceability to the destination. Each aspect of traceability involved adds to the complexity and cost of the program.

Traceability is important when product or process errors are investigated. It may be important to determine which lot of a component was used in a defective or failure-prone product, or which batch of a chemical was used when process results are poor, or who the contractor was that supplied the part. Traceability is also important if product recalls are necessary. There the traceability is to the end user.

As with identification, the concept of traceability also extends to programmatic type items such as procedures, documents, and activities. For important operating instructions, it may be appropriate to know when a specific revision was current and being used. For training, it may be appropriate to know when a specific subject was covered. For inspections and audits, it is important to know who, when, and how the activity was performed. For information, it may be appropriate to know when some data were obtained and who they were obtained from.

To succeed, traceability programs need to be well planned, well implemented, and completely and accurately documented. Arnold states that traceability programs often fail because of lack of planning, discipline, or consistency.[21]

ISO requires the establishment and maintenance of procedures to suitably identify the product from receipt, through production, to delivery and installation. If traceability is specified as a requirement, there should be procedures to uniquely identify the product and record the information.

In implementing product identification and traceability requirements, it is important to decide on appropriate traceability procedures (including those for individual batches); consider the various types of traceability/identification required such as labeling and bar codes; and determine requirements for records retention, such as: availability, retention times, and responsibility.

For safety activities and programs, identification also refers to those kinds of identifications required to meet safety program and regulatory requirements. These would include such things as materials, signs, hazards, and exits. Also included would be unique identifiers for required tests and inspections, and safety, process, and production equipment. Identification also refers to the "soft" items (such as procedures and software) that are used in the safety program.

For safety, it is important to establish customer and/or regulatory requirements for safety identification including the right-to-know; assure that all hazardous materials are properly identified and immediately identifiable; determine the need for postings, signs, warnings, color codings, etc.; and develop unique identification for all safety-related equipment and activities.

PRODUCT IDENTIFICATION AND TRACEABILITY: ASSESSMENT QUESTIONS

- Are processes, materials, products, and services suitably (and uniquely) identified so that safety management objectives can be achieved?
- Is all safety-related equipment uniquely identified?
- Have we determined the types of traceability/identification required: paper vs. electronic, labeling, bar codes?
- Have we determined the following for safety records: availability, retention times, responsibility?
- Do we ensure proper identification of hazardous materials?
- Is there an up-to-date chemical inventory system?
- Do we uniquely identify containers, piping, material conveyors, etc.?
- Have we systematically determined the need for safety signs, warnings, color codings, etc.?
- Are all safety-related inspections, audits, tests, surveys, etc. identified by the place, time, personnel, and equipment used?

IX. PROCESS CONTROL

Process control means that processes are planned, executed, and controlled such that the equipment, environment, personnel, documentation, and material employed constantly result in meeting quality or safety requirements. Process control requires process identification, process planning, and process maintenance. The activities that require control depend on the type of process involved and the inherent risk. In a manufacturing operation, the activities could be cutting, welding, grinding, and painting. In engineering, the activities could include drafting, designing, analyzing, and testing.[22]

A. ISO REQUIREMENTS

ISO requires the identification and planning of production, installation, and servicing processes which can affect quality, and the assurance that these processes are controlled. Process control is exercised through procedures; suitable materials, equipment, personnel, supplies, and working environment; compliance with reference standards, codes, quality plans, and procedures; monitoring and control of process parameters and product characteristics; approval of processes and equipment; clear criteria for workmanship; product standards; and suitable equipment maintenance.

ISO-9004 recommends that process planning include planning for controlled production, planning for quality verification, and planning for in-process and final inspections. Planning for production would include the definition of all controlled conditions, process capability studies, and instructions to ensure satisfactory work completion. The planning of process verification at important stages in the operation may include process control charting and sampling, or appropriate product inspection and testing. Planning for all process testing should include requirements for documentation, procedures, specialized equipment, and workmanship standards.

ISO-9004 also speaks to process capability verification, control of supplies, utilities, and environments, and equipment control and maintenance. All quality-important supplies, utilities, and environments should be controlled and periodically verified to ensure process uniformity. All process equipment, including computers and software, should be well maintained, properly stored, if necessary, and periodically verified (recalibrated) to ensure capability, bias, and precision. A preventive maintenance program is recommended for important equipment.

There also needs to be defined authority and responsibility for process change control. All changes to processes should be documented and the effects of the changes evaluated afterward.

B. IMPLEMENTATION

In obtaining process control, all the factors that affect the process, such as equipment and work environment, need to be defined. Also, all the factors which affect the required product or activity characteristics, such as specifications, workmanship standards, regulatory standards and codes, should be identified and documented. Workmanship standards should be clearly defined or illustrated, using examples if required.

C. SAFETY IMPLICATIONS

For safety management purposes, in addition to chemical and manufacturing and production processes, the meaning of a process is expanded to also include work activities (since they can result in accidents), safety-related activities such as operations, testing, maintenance, training, inspection, and surveillance, and all safety support activities such as emergency planning. Safety-related administrative processes are also included.

For all processes, controls should be in place to help ensure that all activities are carried out safely and correctly, and that all supportive materials and equipment are available. In addition, safe and effective process control requires that any documented process hazards analysis should be reviewed; predictive and preventive maintenance procedures should be developed for monitoring and control equipment; and verification methods for safety-related activities should be developed. For complete safety management of processes, the OSHA Process Safety Management guidelines which apply to the management and control of hazardous chemical processes should be used to the appropriate extent. These guidelines are described and discussed in Chapter 5. They contain material on:

- providing complete information about process hazards, equipment, and technologies;
- completing a process hazards analysis to identify, evaluate, and control the process hazards;
- providing clear and complete process operating instructions;
- effectively training employees to operate the process;
- doing a pre-startup safety review of new and modified processes;
- instituting a complete mechanical integrity (inspection, testing, and maintenance) program for important process equipment;
- requiring specific authorizations for nonroutine work;
- carefully reviewing all process changes and managing how changes are made;
- investigating all significant incidents for necessary preventive and corrective actions;
- planning for potential emergencies; and
- auditing the whole process safety management system.

For effective management and control of hazards, the OSHA Voluntary Protection Program guidelines should be used. These guidelines are described in Chapter 4. In general, process hazards should be controlled (in order of effectiveness) by:

- avoiding the hazards entirely;
- reducing the hazard by substituting less hazardous materials or activities;
- providing engineering controls to prevent accidents or reduce their severity;
- using personal protective equipment to shield workers; or
- instituting administrative controls to prevent reaching dangerous conditions.

D. SPECIAL PROCESSES

It is also important that special processes be identified for suitable monitoring and control. Special processes are those where final inspection and testing may not be enough to verify product quality, and include such processes as plating, welding, and heat treating. These special processes may require comprehensive measurement assurance and equipment calibration. They may also require prequalification, statistical process control, special environmental controls, and special training and certification of staff. Many special processes, such as welding on pressure vessels, have guidance provided by industry standards.

E. PROCESS MONITORING

Process variables need to be monitored, controlled, and verified at specified frequencies to assure: the accuracy and variability of equipment; the skill, capability, and knowledge of workers; the accuracy of measurement results and control data; the requirements for special environments, time, temperature, and other factors; and personnel, process, and equipment certification records.

Verification of activities need not be accomplished by inspection, but can be ensured by proper qualification of personnel, equipment, and methods, and some form of a monitoring or checking process. Even a customer response questionnaire is a form of checking.

PROCESS CONTROL: ASSESSMENT QUESTIONS

- Are processes planned, executed, and controlled such that the equipment, environment, personnel, documentation, and material employed constantly result in meeting safety management objectives?
- Have we established control and verification requirements for important safety-related activities?
- Are there suitable administrative, engineering, and protective equipment controls for all work processes?
- Have we identified critical control points and the factors affecting key processes such as equipment and work environment?
- Have we clearly identified the following product requirements: specifications, workmanship standards, regulatory standards, and codes?
- Have we ensured there are documented procedures for monitoring conformance and control of suitable process parameters and product characteristics during production, installation, and servicing?
- Are all special processes identified, and suitably controlled, and are process personnel qualified?

X. INSPECTION AND TESTING

In ISO-9000, inspection and testing means that materials, products, or services received or produced, and equipment and systems used are verified to meet requirements prior to use, processing, or dispatch. Inspection and testing does not create quality, it only verifies conformance. Their roles are verification, validation, and data gathering. In practice, inspection and test procedures should be designed and maintained to detect problems as early as possible. The critical stages are at input receiving and inspection, crucial process points, and output (final inspection).

A. ISO REQUIREMENTS

ISO requires inspection and testing to assure that specified product requirements are met. This is accomplished through receiving inspection and testing, in-process inspection and testing, and final inspection and testing. Receiving inspection requirements should be based on the amount of subcontractor control, and all received materials should have a clear identification of status, such as released, rejected, quarantined, not tested. Final inspection procedures should assure that no product is released until all quality requirements are met and the data are satisfactory. This also means that all previous tests were satisfactory and documented as such.

B. IMPLEMENTATION

Implementation of an inspection and testing process requires that:

- a general use policy such as, do not use until verified, is established;
- all quality characteristics subject to test are listed;
- approved, current, and complete procedures are available at the inspection and test stations;
- the responsibilities of all inspection personnel are made clear;
- documentation requirements are set; and
- material is only released to the next station when all tests and records are complete.

In traditional quality programs, inspection and testing is a reactive process, applying only to products leaving the door. Defective or nonconforming items are identified, the product is removed and nothing further is done to understand the cause or remedy the situation. In modern programs, the thrust of inspection and testing is proactive. Inspection and testing should gather data for effective corrective action and analysis and for verification of correction.

C. INSPECTION TYPES

Receiving inspection simply verifies that the incoming product meets specifications. It is important that consideration extend not only to purchasing specifications but also to actual use specifications. Mistakes in purchase specifications such as incorrect drawings, could result in obtaining materials meeting specified requirements but not suitable for the process.

Another form of receiving inspection is the source inspection, where conditions at the supplier's facility are inspected. If the supplier's quality and production operations are verified to be acceptable, the amount of actual receiving inspection conducted can be decreased or even eliminated.

In-process inspections ensure conformity at appropriate process points. In-process inspections include: set-up and first piece inspections, inspection or test by machine operator, automatic inspection or test, fixed inspection stations throughout the process, and patrol inspections by inspectors monitoring specified operations. In a job shop environment, in-process inspection ensures that proper drawings or procedures are used and that qualified individuals do the work. In an assembly line environment, sampling or check inspections are performed to verify the process capability.

Final inspections ensure conformity of the finished product. Final inspections focus on the product or on the process and include acceptance inspections or tests, such as screening and lot sampling or product quality auditing of representative units. Final inspection verifies that all receipt and in-process inspections have been accomplished, all final tests or inspections completed, and material is not shipped unless everything checks out satisfactorily

D. DOCUMENTATION

Inspection and test records must be maintained to document that all inspection and test requirements were met. In manufacturing operations, such records may be inspection results marked on a procedure or special test reports on sample products. In a service operation, records might include documented activity verifications.

E. SAFETY IMPLICATIONS

For safety, inspection and testing applies to facilities, and safety-related equipment and activities (also see process control). Regarding equipment, the full range of safety equipment should be included in an inspection and testing program, including: fume hoods, fire protection devices, personnel protective equipment, alarms, machine guards, interlocks, etc. Regarding activities, all the safety inspection and surveillance activities should be planned for, controlled, and documented. These would include such things as normal fire and occupational safety inspections and air quality or noise level surveillances. Specific requirements for safety inspections are given in:

- Chapter 4, IX: Baseline Surveys,
- Chapter 4, X: Site Inspections,
- Chapter 4, XIII, Facility Equipment and Maintenance, and in
- Chapter 5, IX: Mechanical Integrity.

Finally, in a proactive safety program, inspection and testing should only be seen as validation. For example, if a routine inspection shows several portable ladders to be damaged, this should be viewed as a programmatic failure - the damage should really have been seen, identified, and reported by the previous users. The philosophy of safety through inspections should not exist.

INSPECTION AND TESTING: ASSESSMENT QUESTIONS

- Are safety-related materials, products, or services received, safety equipment and systems used, and products, or services provided, all verified to meet safety requirements prior to use?
- Do we have procedures for receiving inspection and testing, operability inspection and testing, in-process inspection and testing, and final inspection and testing?
- Do we have a policy of "do not use until verified"?
- Have we identified categories of product and listed all safety management characteristics subject to inspection and test?
- Have we identified all safety equipment requiring testing?
- On receipt, do we provide for clear status identification, such as released, rejected, quarantined, not tested?
- Is the inspection and testing system proactive, with data used for effective analysis and corrective action?
- Is there a routine and complete facility safety inspection program conducted by trained personnel?

XI. CONTROL OF INSPECTION, MEASUREMENT, AND TEST EQUIPMENT

Control of inspection, measuring, and test equipment means that devices, techniques, and reference standards used to verify that products or services meet quality requirements are effectively controlled, calibrated, maintained, and used. This control also extends to test software. For safety, this ISO requirement means that all inspection, measurement, and test equipment used in a safety-related process or in an activity with safety implications should be subject to a defined control, calibration, and maintenance program.

A. ISO REQUIREMENTS

ISO requires: measurement and accuracy identification; measurement equipment identification and calibration; calibration procedures; ensuring acceptable accuracy and precision; test equipment and calibration status identification; assessment of out-of-calibration results; proper environmental conditions; proper equipment handling; and proper safeguarding of the facilities used. ISO also requires that test equipment is used in a manner which ensures that its measurement uncertainty is known and is consistent with any required measurement capability. This means that procedures must contain specific requirements regarding proper use and conditions of use.

Test software or hardware should be checked to ensure that they verify the acceptability of product prior to actual use, during production, installation, or servicing and at periodic intervals thereafter.

B. IMPLEMENTATION

Implementing an effective control of an inspection, measurement, and test equipment program requires:

- identifying all inspection and test requirements, including accuracy requirements;
- listing all equipment required to conduct inspections and tests, such as:
  - laboratory equipment,
  - inspection and test equipment,
  - production machinery,
  - jigs, fixtures, templates, patterns, and related software;
- identifying recognized calibration requirements for all fixed and portable equipment;
- ensuring that there are procedures and documentation for the following:
  - measurements to be made,
  - calibration procedures,
  - measurement uncertainty,
  - identification of equipment calibration status,
  - out of calibration action,
  - environmental control, handling and storage, and safeguarding against unauthorized adjustment, and
  - rechecking intervals;
- establishing an effective record system documenting all equipment, calibrations, and corrective actions;
- ensuring the calibration of equipment used for design verification and validation;
- effectively controlling the facilities and environments for: calibration, handling, control, storage, and maintenance of all necessary measuring and test equipment and standards;
- suitably identifying all inspection equipment showing calibration status;
- ensuring that the accuracies of all standards used for calibrations are traceable to national or international standards; and
- providing appropriate adjustment, repair or recalibration, where necessary.

ISO-9004 also recommends that the measurement process itself be monitored and maintained under statistical control. Measurement errors should be compared with requirements and corrective action taken as needed.

C. CALIBRATION

The measuring equipment used must be appropriate, capable of the required accuracy and precision, and handled to maintain its integrity. The equipment and facilities should be protected. The equipment should be calibrated at periodic intervals or before use, and the calibration results and status should be identified and documented. The reference standards used should be traceable to national or international standards or to special documented criteria.

Similarly, the quality of any laboratories used for sample analysis should be determined prior to using them. The standards used by these laboratories should be traceable, the procedures should be technically correct, the history of past performance should be excellent, and there should be an internal quality control program that ensures that all activities are carried out as specified and correctly. A common method of checking the performance of outside laboratories is to use split or spiked samples where other laboratories also determine the results or a known sample is sent for analysis.

Arnold[23] cites that many companies do not understand the need for a calibration program. These companies believe that calibrated test equipment is not needed or that the measurements are not that critical. Both reasons strike at the heart of quality management - for quality in production, processes, or activities can only be controlled, managed, and improved if it is measured.

ISO requires that the calibration program only extends to instruments and equipment required to ensure quality and safety. Instruments only used to verify or check status, for example, need not be calibrated, they only have to be operable to do what is desired. However used though, all the instruments and their allowed uses should be suitably identified.

D. CORRECTIVE ACTION

Finding processes out-of-control or equipment out-of-calibration should lead to appropriate corrective action. For out-of-calibration equipment, the effect on completed work should be assessed and the extent of recall, reprocessing, retesting, or rejection determined. In addition, the cause of the calibration failure should be determined and corrected, including modifying the calibration program if necessary.

E. PROCEDURES

Inspection procedures and documents should be as detailed as necessary to provide adequate control, analysis, and documentation. Equipment calibration procedures should contain:

- the type of equipment;
- the identification or serial number;
- the equipment location;
- the calibration frequency;
- the calibration method;
- the specific acceptance criteria; and
- the actions to be taken if the equipment is out of calibration.

Calibration procedures can be specific or general, applying to a particular component or an entire class of equipment.

CONTROL OF INSPECTION, MEASUREMENT, AND TEST EQUIPMENT: ASSESSMENT QUESTIONS

- Are all devices, techniques, and reference standards which verify that safety requirements are met: controlled, calibrated, maintained, and stored properly?
- Have we identified all inspection and test requirements including accuracies?
- Have we listed all equipment required to conduct inspections and tests, such as: laboratory equipment, inspection and test equipment, surveillance equipment, production machinery, jigs, fixtures, templates, and software?
- Have we identified recognized calibration requirements for each piece of equipment, either fixed or portable?
- Are there remeasurement or verification procedures when equipment is found out-of-calibration?
- Is all equipment marked with calibration status?
- Are calibrations traceable to national, international, or appropriately documented standards?

XII. INSPECTION AND TEST STATUS

Inspection and test status means that products, processes, or equipment are identified such that any uninspected or nonconforming items or activities are distinguished from inspected and safe and conforming items or activities. The identification should ensure that only products, processes, or equipment that have passed the required inspections are used, dispatched, or installed. ISO lists examples of types of status identification such as: markings, authorized stamps, tags, labels, routing cards, inspection records, software, or physical location.

A. IMPLEMENTATION

Implementation requires: identifying locations where inspection status is critical; determining the means of identification and status; establishing the responsibility and authority for release; providing a means for the controlled use or release, by authorization, of materials, products, or equipment that has not passed the required inspection and tests; and ensuring that inspection and test status are part of the traceability record, if required.

The areas where inspection status is critical include: receiving, production, post production, installation, and servicing.

B. RECEIPT STATUS

The receipt of new materials and equipment is one common example of inspection and test status in action. Usually, new materials and equipment are inspected for acceptance on receipt, then moved to another location for eventual use or delivery. Large items may be individually marked as inspected and accepted while smaller items are commonly just stored with other acceptable items for use. Any nonconforming items are segregated for further action. In this case, the physical location of the item indicates its inspection and test status.

C. SAFETY IMPLICATIONS

All safety-related materials, products, or equipment should be identified to verify conformance or nonconformance with regard to tests and inspections required either by regulation or by safety program procedure. The location and inspection status of each such item should be identified, such that the status is clear to all potentially affected personnel. Very common identification examples are inspection tags on fire extinguishers and calibration stickers on pressure gauges.

In parallel with this requirement, administrative controls should be in place to ensure that equipment not identified as conforming with safety/test requirements is not used. For more details, see the next section on control of nonconforming product.

INSPECTION AND TEST STATUS: ASSESSMENT QUESTIONS

- Are products, processes, or equipment identified such that any uninspected or nonconforming items or activities are distinguished from safe and conforming items or activities?
- Have locations where inspection status is critical, such as: receiving, production, post production, installation been identified?
- Have the means of identification/status been determined including: marking, stamps, tags, labels, routing cards, hard copy vs. electronic records, and physical location?
- Have all positive release procedures and responsibilities been reviewed?
- How do we ensure that all safety-related materials, products, or equipment are identified to verify conformance or nonconformance with regard to tests and inspections?

XIII. CONTROL OF NONCONFORMING PRODUCT

Control of nonconforming product means that products, processes, or equipment identified as nonconforming or unsafe are prevented from inadvertent use, and that measures are used to control remedial action to make such items or processes acceptable. In ISO-9000, the purpose of this section is to prevent nonconforming products from reaching the customer. In terms of safety, the meaning of this section is broadened to include all identified defects and deficiencies in materials, activities and operations that affect safety. The safety goal is to ensure that work safety is not compromised by the identified defects and deficiencies.

A. ISO REQUIREMENTS

The control of nonconforming product means that there must be procedures for: identification, documentation, segregation (as practical), disposition, and notification of all functions impacted. ISO also requires assigning responsibility and authority for disposition approval where disposition means rework, accept, regrade, or reject.

ISO-9004 recommends the immediate identification and recording of nonconformances. It is important that those who review nonconformances are competent to do so and to determine further use or disposition. Reviews of nonconformances should consider the full range of potential effects. These include effects on interchangeability, further processing, performance, reliability, safety, and esthetics.

B. SAFETY IMPLICATIONS

For a continuously improving safety management system, it is vital that procedures be developed to determine the root causes of nonconformances. Possible root causes include: defective engineering controls, nonconforming materials, and poor training.

Typical nonconformances would include:

- procedures that are ineffective, have errors, or don't correspond to existing conditions;
- equipment and machinery that is defective, or does not meet approved safety standards or engineering requirements;
- safety systems that are degraded due to equipment or test failure, or because equipment is out-of-service for testing or maintenance, or because proper documentation does not exist;
- safety program elements that are suspect because of repeated failures, accidents, operator mistakes, or audit or inspection findings;
- safety management problems related to poor or ineffective performance.

If a safety concern is identified and upon review the activity or process is allowed to continue, or the equipment is allowed to be used, further action must be taken to ensure that conditions remain safe. In this situation, all affected parties should be made aware of the concern and what is being done to prevent harm. A common safety example is the use of a fire watch in situations where an alarm system is degraded or part of a fire protection system is disabled.

CONTROL OF NONCONFORMING PRODUCT: ASSESSMENT QUESTIONS

- Are products, processes, or equipment identified as unsafe or nonconforming prevented from inadvertent use and are measures taken to control remedial action to make the items safe?
- For nonconforming products, processes, or equipment, have procedures for: identification, documentation, segregation, prevention of inadvertent use or installation been developed and implemented?
- Have procedures to determine a nonconformance root cause, to report it to management, and to ensure corrective action, been developed and implemented?
- Are appropriate compensatory actions used to protect against temporary hazards such as fire protection impairments?

XIV. CORRECTIVE AND PREVENTIVE ACTION

Corrective and preventive action means that effective action is taken to prevent the occurrence and recurrence of nonconformities (unsafe items, processes, or equipment). Corrective action is reactive, directed to eliminating causes of actual nonconformities, while preventive action is proactive, directed to eliminating causes of potential nonconformities. Continuous improvement is the result of identifying and correcting the causes of nonconformances.

Corrective action programs are often weak, addressing only the immediate problems, while failing to avoid recurrences. Another weakness is that corrective action programs often do not address procedures or systems, but focus only on products or services. One feature of stronger programs is having a corrective action log documenting and tracking the status of corrective action requests, actions, and follow-ups.

It is important that adequate resources are provided for full and effective implementation of preventive and corrective action activities. Such resources include: expertise on the part of the analysis and review staff; records of all tests, audits, and problems; records of all safety suggestions and recommendations; instruction procedures; access to proper testing and analysis equipment; and examples of defective product for analysis. In addition, preventive action requires the availability of a level of professional, operational, and analysis knowledge greater than that needed to just analyze and correct identified problems.

A. SAFETY IMPLICATIONS

For safety management, this section refers to any safety problems or events. These could be: equipment, process, program or human problems; test or surveillance failures; or actual incidents or accidents. For example, the corrective action process could be applied to a critical pump failure, or to the finding of higher than normal toxic material concentration levels, or to the accident which caused some worker's broken leg.

B. CORRECTIVE ACTION STEPS

The steps in the corrective action process are: assigning responsibility and authority; identifying nonconformances; segregating nonconforming materials; evaluating the impacts of the nonconformances on quality and safety; investigating to determine the root cause; instituting corrective and/or preventive action; instituting process controls to ensure correction; and documenting the whole process including changes to procedures and methods.[24] In addition the corrective and preventive action process should be reviewed to determine if it is indeed effective in correcting and preventing problems. There should be procedures for documenting the corrective and preventive action process and tracking the results of actions to verify effectiveness. All relevant information on actions taken should be submitted for management review.

ISO-9004 specifies that responsibility and authority for the coordination, recording, and monitoring of corrective action should be assigned to a specific function in the organization. However, the actual actions involved may be conducted by any individual or group.

C. PREVENTIVE ACTION

ISO recommends that all appropriate sources of information should be used to investigate the need for preventive action. Among those sources listed are: audits, quality records, service reports, and customer complaints. For safety, the whole range of safety-related information should be investigated. This would include reports of near-miss accidents, instances of inappropriate behaviors, and worker suggestions. Important also are sources from outside the organization such as national accident statistics, reports of events at other facilities, and regulatory guidance. For processes, the techniques of process hazards analysis should be used (see Chapter 5 in this text) to evaluate the need for preventive action.

CORRECTIVE AND PREVENTIVE ACTION: ASSESSMENT QUESTIONS

- Is effective action taken to prevent the occurrence and recurrence of unsafe items, processes, or equipment?
- Have procedures for corrective action (actual accidents/injuries) and preventive action (potential accidents/injuries) been developed?
- Do the procedures cover initiating corrective actions? verifying effectiveness? and documenting any necessary changes?
- Are accidents/injuries immediately reported to high level management (the CEO)?
- Are senior level executives or only front-line managers involved in accident investigations?
- Are there procedures to analyze all processes, work operations, concessions, safety records, failure and incident reports, near-miss accidents, audit/inspection findings, safety-related recommendations and suggestions, and complaints to detect and eliminate accident and injury causes?
- Do accident investigations seek to determine the accidents' root cause?
- Are corrective and preventive actions appropriate to the magnitude of the problem?
- Are the corrective and preventive actions effective?
- Are professional safety and health personnel employed in the preventive action analyses?
- Do accidents, injuries, events not repeat past occurrences?
- Do the causes of accidents, injuries, and events not repeat?

XV. HANDLING, STORAGE, PACKAGING, PRESERVATION, AND DELIVERY

Handling, storage, packaging, preservation, and delivery means that there are measures to: identify the hazards related to the specified activities; ensure that materials, equipment, and products are not so affected as to impact worker safety; and ensure that there are appropriate safety procedures and training to conduct the activities. In ISO-9000, the intent of this section is to focus on the materials and products, to ensure that they are handled properly and will be delivered in good condition to the customer. The requirements apply to all stages of production and handling where the activities of handling, storage, packaging, preservation, and delivery might take place.

A. ISO REQUIREMENTS
Specifically, for handling, ISO requires methods and means to prevent damage or deterioration. For storage, ISO wants secure storage areas to prevent damage or deterioration, with appropriate means of authorizing receipt and dispatch. For packaging, ISO seeks control to ensure conformance to specified requirements. For preservation, ISO wants appropriate methods of preservation and segregation of product. And for delivery, ISO seeks protection of the product after final inspection, which could include delivery if specified by the contract.

Handling and storage procedures should ensure the use of correct pallets, containers, conveyors, and vehicles. Damage from vibration, shock, abrasion, corrosion, and temperature should be prevented. Stored items should be checked periodically. Shelf-lives should be carefully adhered to as well as proper environmental conditions. All materials should be legibly marked from delivery to final destination, to protect against loss or the use of the wrong materials or items. Packaging procedures should cover packing details including moisture elimination, cushioning, and blocking and crating. Packaging materials should also be suitably controlled. Delivery instructions should include provision for the identification of materials with short shelf-lives or special handling and protection requirements.

B. IMPLEMENTATION

Implementation requires: reviewing all the material and product handling processes to identify critical points; reviewing available information (damage rates, etc.); generating and revising procedures, instructions, and documentation for packaging designs, in-process handling, warehousing, transportation techniques and carrier selection, storage, preservation and segregation methods, and environmental impact; ensuring that procedures prescribe methods to prevent damage or deterioration; ensuring that proper safety methods such as ergonomics, grounding, stacking heights, safe-handling instructions, environmental protection equipment, and engineering controls are used; ensuring that the receipt and release from stock rooms and/or other storage areas are controlled; ensuring that there are procedures for the identification, segregation, and the assessment of condition at defined intervals; ensuring that procedures cover protection and quality after final inspection and during transit to final destination; ensuring that there are procedures to safely control the full range of material handling activities; and ensuring that personnel are trained and qualified to perform material handling duties.

C. SAFETY IMPLICATIONS

For safety management, the intent is to focus on the hazards associated with materials and all of the activities used to handle, store, and transport materials and products. For example, chemicals should be prevented from inadvertent mixing or contamination, or being subject to excessive temperatures or environmental conditions that could cause deterioration. Equipment should be protected from mechanical damage, water and humidity damage, temperature damage, corrosion, and contamination. Materials with limited shelf-lives or special storage requirements should be identified and stored and handled appropriately. Delivery would include requirements for fleet and transportation safety.

HANDLING, STORAGE, PACKAGING, PRESERVATION, AND DELIVERY: ASSESSMENT QUESTIONS

- Do we have measures to identify the hazards related to activities, and ensure that materials, equipment, and products are not so affected as to impact worker safety?
- Have all processes been reviewed and critical points identified using safety information?
- Have procedures for handling, storage, packaging, preservation, and delivery been reviewed, revised, and improved and have the revisions been evaluated?
- Have we ensured that proper safety methods such as ergonomics, grounding, stacking heights, safe-handling instructions, environmental protection equipment, and engineering controls are used?
- Are all packages properly marked noting contents, handling requirements, and hazards.
- Are storage areas adequate and environmentally controlled?
- Is access to storage areas controlled to prevent product damage or personal harm?
- Are there specific packaging requirements for potentially dangerous or hazardous materials?

XVI. CONTROL OF QUALITY RECORDS

Control of quality records means that records are established, documented, and maintained to demonstrate achievement of quality or safety requirements, including regulatory requirements, and quality or safety system effectiveness. To be useful, records must be controlled, accessible, and identifiable. In ISO-9000, the intent of quality records is to verify that the quality program is conducted to requirements. Records also verify the condition of the product and provide a history of the part, process, or program.[25]

A. ISO REQUIREMENTS

ISO specifically requires procedures for identification, collection, indexing, filing, storage, maintenance, and disposition of records including pertinent subcontractor records. Records must be legible, stored, and readily retrievable, and retained for some established time period. Storage facilities must have a suitable environment to prevent damage, loss, or deterioration. Records may be maintained in any media, such as hard copy or electronic.

B. IMPLEMENTATION

Records control implementation requires reviewing the current list of documents and reviewing procedures to identify, collect, index, file, store, maintain, and dispose of records. For each category of records, the following issues and requirements: legibility, identification with a product, ability to be retained, storage environment, retention needs, and availability to purchaser, must be addressed.

C. QUALITY RECORDS

For safety and quality programs the types of records involved can be very extensive. ISO specifically requires the retention of the following types of quality records: management reviews; contract reviews; design reviews; design verification measures; acceptable suppliers; customer-supplied product that is damaged; product identification and traceability; qualified processes, equipment and personnel; product released for urgent production; inspection and test records; test hardware and software checks; calibration records for test equipment; review and disposition of nonconforming product (when required by contract); corrective action investigations; internal quality audit results; internal audit follow-up activities; and training records.[26] Other records could include: service reports, trip reports, and qualification records. Computer records and files should be periodically backed-up and stored in a safe location.

D. SAFETY RECORDS:

For safety management, records are necessary for regulatory compliance, for ensuring program implementation, and for providing a basis for continuous improvement through review and analysis. Effective accident investigation and analysis also require a well-documented history of activity, inspection, and statistical records. Safety management records include the following:

- accidents (injuries, production interruptions, product damage, facility damage);
- near-miss accidents;
- safety and health training;
- special hazards exposures;
- safety meetings;
- medical evaluations and histories;
- safety inspections, audit and review records;
- emergency plans and drills;
- hazardous chemical training, communication, and monitoring;
- safety, fire prevention and emergency planning procedures;
- materials handling training, certification, and procedures;
- noise conservation program documents; etc.

CONTROL OF QUALITY RECORDS: ASSESSMENT QUESTIONS

- Has there been a review to determine requirements for safety management records throughout the system?
- Are records established, documented, and maintained to demonstrate achievement of safety management system objectives and to meet regulatory requirements?
- For each category (function) of documents, have procedures been established to: identify, access, collect, index, file, film, store, maintain, retain, and dispose of the records?
- Are the control and retention requirements for the following types of documents included: OSHA 200 logs, safety audit reports, accident investigations, safety device inspections, maintenance, and corrective action reports for safety items?
- Are safety records readily retrievable?
- Are safety records stored in an environment that prevents damage and deterioration?

XVII. INTERNAL QUALITY AUDITS

Internal quality audits means that comprehensive audits are planned and executed to verify the existence, implementation, and the effectiveness of the quality or safety management system. All activities and operations should conform to requirements, and the system should meet specified goals and objectives. The intent of auditing is not only to verify compliance but also to promote continuous improvement through the identification of weaknesses and the presentation of corrective recommendations. Meeting this intent requires the presence of highly qualified audit personnel.

ISO expects audits to be carried out on a schedule consistent with the importance of the activity. Also, personnel conducting the audit should be independent of the area being audited. Once deficiencies are brought to the attention of management, there should be timely corrective action taken.

According to ISO-9004, audit evaluations could include: organizational structures; administrative and operational procedures; personnel, equipment, and resources; work areas, operations, and processes; items being produced or activities being conducted; and documentation, reports, and record keeping. Audit reports should include specific examples of deficiencies or noncompliances, possible reasons for the deficiencies, recommendations for corrective action, and a review of how effective past corrective actions were implemented.

A. IMPLEMENTATION

Implementation of an audit program requires: identifying the activities to be audited; establishing the qualifications of audit personnel, including their experience, training, availability, and knowledge of auditing techniques and the safety program and hazards; developing or updating audit procedures to include planning, conducting, documenting, and reporting findings, conclusions, and recommendations; conducting an initial (trail) audit to evaluate the adequacy of procedures, determine their implementation and effectiveness, verify compliance with requirements and the meeting of program objectives; and establishing an extended (at least annual) audit program including scheduling audit frequency based on the importance of the activity, documenting any follow-up, reporting to management, and initiating timely corrective action.

The audit program should also ensure that corrective actions fix the problems in products, processes, and systems or create change when improvements are needed. Evidence should exist to show that:

- audits meeting specified requirements have been implemented;

- activities are audited by personnel independent of those having direct responsibility for the activity being audited; and
- follow-up audits verify the timely implementation and effectiveness of corrective actions. Audits should always include interviews of key personnel and managers.

A discussion of the audit process and ISO auditing requirements is given in Appendix B, Auditing.

B. SAFETY AUDITS

The range of safety program audits includes: visual inspections of conditions and activities, documentation audits for compliance and completeness, and comprehensive program audits or reviews. The OSHA VPP guidelines (see Chapter 4, VII) recommend a review (a comprehensive audit) of safety program operations at least annually to evaluate their success in meeting safety goals and objectives. A comprehensive program audit is essential periodically for the evaluation of the whole set of safety and health management means, methods, and processes, to ensure that they are adequate to protect against the potential work-site hazards. The audit determines whether policies and procedures are implemented as planned and whether in practice they have met the objectives set for the program. It also determines whether the objectives provide sufficient challenge to lead the organization to meet the program goal of effective safety and health protection. When either performance or the objectives themselves are found inadequate, revisions are made. Without such a comprehensive review, program flaws and their interrelationship may never be caught and corrected.

INTERNAL QUALITY AUDITS: ASSESSMENT QUESTIONS

- Are comprehensive safety program audits and reviews planned and executed to verify the effectiveness of the safety management system?
- Is the entire program audited on at least an annual basis?
- Have the activities to be audited been identified and do they include the whole set of safety and health management means, methods, and processes, to ensure that they are adequate to protect against the potential hazards at the work-site?
- Are audit priorities based on risk, previous problems, and audit results?
- Do the audits determine whether policies and procedures are implemented and are meeting program objectives?
- Have the qualifications of audit personnel been established, including the following: experience, training, knowledge of safety program and hazards, and availability?
- Are auditors independent of the area being audited?
- Do audit reports list deficiencies, reasons, and corrective recommendations?
- Are the responsible managers accountable for time and appropriate corrective action?
- Is timely corrective action taken to correct audit findings?
- Are previous corrective actions followed-up?

XVIII. QUALITY TRAINING

Quality training (just "training" in the ISO standards) means that the training needs of personnel whose work affects quality and safety are identified, and that training is provided so that the personnel are qualified to do their jobs. For specific tasks, qualification can be based on appropriate education, training, and/or experience. Appropriate training records should be maintained, and personnel performing the training should be adequately qualified.

A. IMPLEMENTATION

Peach[27] describes the implementation requirements for a training program as:

- identifying the training and experience needs;
- listing all job functions, and establishing training and experience requirements for each function, including the requirements in job descriptions;
- providing training based on the following: quality (safety) plan elements, process knowledge requirements (methods and equipment), product and activity knowledge requirements (specifications, workmanship standards), cross-training, other requirements;
- establishing and recording personnel qualifications in individual files to include completion of all required training, education (initial and additional), previous experience, physical characteristics and limitations, special training, medical records, awards, rewards, promotions, and cross-training;
- and developing a training plan (matrix) to include required training, optional additional training, periodic reevaluation.

B. ISO RECOMMENDATIONS

ISO-9004 recommends that the training needs of executives and managers, technical personnel, and production supervisors and employees should be considered. Executives and managers should understand the quality system and tools and should be able to evaluate quality system effectiveness. Technical personnel should receive training to enhance their contributions and production personnel should be trained for full performance. Requirements for the certification of personnel performing specialized activities should be evaluated and instituted as necessary.

The training program should also:

- ensure that training records are available showing the safety skills training received by each employee including: when and by whom skills were imparted, how trained, how on-the-job training was assessed, and an authorized sign-off on satisfactory accomplishment;
- ensure that training effectiveness is determined by a variety of standard means including: (a) retention (immediately after the training), (b) knowledge (by testing), (c) behavior (by observation and self-reporting), and (d) results;
- use open-ended questions to determine training effectiveness and worker understanding;
- ensure that training records are available to those who might need them on a regular basis, such as supervisors;
- ensure an intensive, first-day indoctrination for new hires;
- ensure that all specific training has been performed or an action plan is in place to fill any gaps;
- and ensure that there are documented methods for the qualification of personnel performing special processes and operations.

For a continuously improving quality system, training should also cover:

- problem solving;
- use of statistical techniques and methods;
- customer feedback; and
- team leadership skills.

C. SAFETY TRAINING

For safety programs, specific training topics include:

- hazard prevention and control including engineering controls;

- safe work procedures;
- administrative controls;
- facility equipment and maintenance;
- emergency planning and preparation;
- the medical program;
- workplace analysis including hazard identification;
- hazard reporting;
- accident/illness investigation and injury analysis;
- safety and health for employees (how to protect themselves and others); and
- responsibilities of supervisors and managers.

Training for emergencies is essential in minimizing the harmful consequences of an accident or other threat if it does occur. Personnel should be thoroughly trained so that their reactions to emergencies are immediate and precise. Otherwise they may expose themselves and others to greater danger rather than reduce their exposure.

For occupational safety, OSHA recommends[28] that a training program contain the following elements:

- determining if training is needed;
- identifying training needs (through OSHA requirements, job and task analysis, accident investigations, employee observations, and review of other programs);
- identifying goals and objectives;
- developing learning activities;
- conducting the training;
- evaluating program effectiveness and improving the program.

In addition, the program should also address periodic refresher training, certification for specific skills, and periodic recertification.

Safety training should be provided based on the degree of hazard faced by the personnel and also by the existence of factors that lead to higher risk. OSHA also identifies the following factors that lead to a high incidence of workplace injuries and illnesses. These factors are:

- employee age (younger employees have more accidents);
- length of time on the job (new employees are more susceptible);
- size of firm (medium-sized firms have more incidents);
- type of work performed; and
- the use of hazardous substances.[29]

Training employees who are at risk places more emphasis on the specific requirements of the job and the possibility of injury. For specific jobs, training may be based on a job hazard analysis which identifies:

- the job description;
- the job location;
- the key job steps in the order performed;
- the tools, machines, and materials used;
- the actual and potential safety hazards associated with the job steps; and
- the safe practices, apparel, and equipment required for each job step.

QUALITY TRAINING: ASSESSMENT QUESTIONS

- Have procedures been established for identifying safety training and experience requirements for all personnel?
- Are the safety training needs of personnel identified and are these personnel properly trained and qualified?
- Have training needs been identified by establishing training requirements for each function?
- Are the training requirements included in job descriptions?
- Are individual files established to record qualifications and the following information:
  - completion of all required training?
  - education (initial and additional)?
  - previous experience?
  - physical characteristics and limitations?
  - special training?
  - medical records?
  - cross-training?
- Are there reliable methods to demonstrate the effectiveness of training?
- Does the training program cover: hazard prevention and control? safe work procedures? administrative controls? facility equipment and maintenance? accident and illness investigation? hazard reporting? and emergency planning?
- Is safety training provided based on the degree of hazard personnel face?
- Does training for specific jobs include the actual and potential hazards associated with each job step and how to protect against them?
- Is the need for additional training considered in all performance appraisals and incident investigations?

XIX. SERVICING

In ISO, servicing means that controls are employed to ensure that servicing operations are carried out to meet requirements specified in the purchasing contract. Servicing may be routine preventive maintenance or it may be corrective maintenance and repair.   Under ISO, warranty repairs are not covered. In terms of safety management, servicing activities should be planned and prepared for to ensure they are done safely.

Implementation requires: identifying purchaser service requirements; documenting servicing requirements to establish procedures, perform the service, and verify that requirements are met; and ensuring that servicing responsibilities are defined for the supplier, distributor, and user.

For safety in servicing it is important that:

- all servicing safety requirements are established and documented including specific procedures, personal protective equipment, road travel restrictions, emergency plans, and installation instructions (including hazard notifications);
- the availability and readiness of special tools or equipment is ensured;
- servicing personnel are properly trained, including driver training and the use of special equipment;
- document control is maintained for spare parts lists, service literature, servicing instructions, and safety instructions; and
- the calibration and recall control of measuring and test equipment used in servicing is maintained.

Servicing is an activity performed at the purchaser's location. As such, servicing personnel should be cognizant of any specific hazards that exist or requirements they might be subject to at the servicing site. Depending on the nature and severity of site hazards, escorts or indoctrination may be provided. Servicing personnel should follow site safety requirements and should exercise extra care, recognizing that their knowledge of site hazards and conditions may not be complete.

Feedback from purchasers is also important in the planning of servicing activities. The history of problems with products or equipment should be used to help determine servicing needs in terms of personnel, supplies, training, and support. Feedback should also be sought regarding the safety performance of servicing personnel. Deficiencies in performance, including not adhering to site safety requirements, require immediate corrective action. The performance of safe servicing activities is the result of effective planning and preparation.

Specific guidance on contractor safety (applicable to servicing) is found in Chapter 4, VI: Contract Workers and in Chapter 5, VII: Contractors. For work on or near hazardous processes, servicing personnel should:

- be informed of known or potential fire, explosion, or toxic release hazards;
- be informed of applicable provisions of the emergency action plan;
- be trained in work practices to safely perform the job; and
- follow the safety rules of the facility.

SERVICING: ASSESSMENT QUESTIONS

- Are controls employed to ensure that any servicing operations are carried out with worker safety in mind?
- Are purchaser servicing requirements identified?
- Are servicing responsibilities defined for the supplier, distributor, and user?
- Are servicing safety requirements established and documented including: specific procedures, personal protective equipment, driver training and road travel restrictions, site hazards, emergency plans, and installation instructions (including hazard notifications)?
- Are servicing personnel appropriately trained?
- Is feedback sought and obtained on the safety performance of servicing personnel?
- Is extra training provided to servicing personnel working at highly hazardous locations?

XX. STATISTICAL TECHNIQUES

Statistical techniques means that measures are taken to control the selection and application of statistical techniques used in controlling processes and operations and determining the effectiveness of meeting process and operational goals. The general objective of using statistical techniques is to remove any external cause of variation in order to produce products in a consistent manner. Statistical techniques should be considered for all stages of the activity or process.

ISO-9004 lists typical statistical methods. These methods are: design of experiments and factoral analysis, analysis of variation and regression analysis, safety evaluation and risk analysis, tests of significance, quality control charts, and statistical sampling inspection. Other statistical analysis and problem solving techniques are described in:

- Chapter 2, X: Statistical Process Control. The typical statistical tools are: pareto charts, run charts, histograms, scatter diagrams, tally charts and various types of control charts.
- Chapter 2, XI: Structural Problem Solving. The tools are pareto diagrams, pareto cause analysis, delphi methods, null-op method, inductive/deductive reasoning, nominal group

technique, fishbone (cause and effect) diagram, why-why diagram, affinity diagram, brainstorming, screening matrix, comparison matrix, boundary analysis, force-field analysis, and quality function deployment.

- Chapter 2, XII: The Best Techniques, or using baselines and benchmarks to assess performance and compare the safety management system with others.

Hazards analysis processes are described in:

- Chapter 5, IV: Process Hazards Analysis. The typical tools are what-if/checklist, hazard and operability study (hazop), failure mode and effect analysis (fmea), fault tree analysis, event-tree analysis, cause-consequence analysis, and human reliability analysis.

In terms of safety management, there are two separate objectives involved with the application of statistical techniques. First, statistical techniques should be used for the proper control of all processes and production activities. Through the application of statistical techniques, products will meet requirements, processes and production activities will be safety conducted, and no hazardous operations will be conducted. Statistical methods are typically used for: product design, reliability specifications, longevity and durability determination, process control and process capability studies, determination of quality levels or inspection plans, data analysis, performance assessment, or defect analysis and process improvement.

Secondly, statistical techniques should be used for the management of the safety program itself. This means that safety performance measures should be developed and statistically analyzed, safety audits should be planned and conducted in a statistically sound manner, and statistical techniques should be used in hazards, accident reliability, and risk analyses.

For all uses of statistical techniques, there should be a review of the status, correctness, and effectiveness of the statistical applications by establishing process capabilities and verifying product (and performance) characteristics. Of course, personnel should be trained and qualified to use the statistical techniques.

All sources of data should be reviewed to determine what data should be used for safety program performance assessment. Typical data sources are:

- internal inspections, audits, and reviews;
- insurance company, OSHA, and other external inspection results;
- accident and injury data including near-misses;
- behavioral observations; and
- insurance loss reports.

Industry-wide data available from OSHA, insurance organizations, journal articles, and other sources can be used to pinpoint typical problem areas and to compare program results. In all cases, care must be exercised to ensure that statistically valid samples are analyzed and the results are statistically meaningful.

STATISTICAL TECHNIQUES: ASSESSMENT QUESTIONS

- Have we selected and applied statistical techniques to control activities and meet safety objectives?
- Have existing statistical applications and procedures been identified and have we reviewed the status, correctness, and effectiveness of the statistical applications?
- Have trained and qualified personnel been provided to apply the statistical techniques?
- Are there documented procedures for implementation, control, and application of statistical techniques?
- Does the range of techniques include: control charts, trend analyses, bar charts, correlating studies, and designed experiments?

XXI. SUMMARY - ISO EVALUATION

ISO - 9000 Topic	Rating*
1. Management Responsibility	
2. Safety Management System	
3. Contracts Review	
4. Design Control	
5. Document and Data Control	
6. Purchasing	
7. Control of Customer-Supplied Product	
8. Product Identification and Traceability	
9. Process Control	
10. Inspection and Testing	
11. Control of Inspection, Measurement, and Test Equipment	
12. Inspection and Test Status	
13. Control of Nonconforming Product	
14. Corrective and Preventive Action	
15. Handling, Storage, Packaging, Preservation, and Delivery	
16. Control of Quality Records	
17. Internal Quality Audits	
18. Quality Training	
19. Servicing	
20. Statistical Techniques	
Total Safety Program Score	

A. ISO EVALUATION RATING GUIDE

(Rate each topic 0 to 4. Consider the scope and breadth of deployment and the results achieved.)

4= Basically all of our people or activities meet the stated criteria or their intent with excellent results achieved.

3= Most of our people/activities meet the criteria or their intent with good results.

2= About half of our people/activities meet the criteria or their intent with positive results.

1= Some or a few of our people/activities meet the criteria or their intent with a few positive results.

0= None of our people/activities meet the criteria and no positive results are evident.

B. ISO SAFETY SYSTEMS SCORING

Total Score   *Safety System Evaluation*

68-80   *ISO quality is completely built into our safety system with very few or no gaps or problems.*

55-67   *Our company has very progressively built ISO quality into safety with sound approaches and few gaps in deployment and integration, but some fine tuning is possible in several areas.*

41-54   *Our company is progressive in building ISO into safety but gaps in deployment exist and refinements are still needed.*

27-40   *Our company has made progress in applying ISO quality to safety but significant gaps in deployment exist. Improvement is needed throughout.*

14-26   *Our company is just beginning to build ISO quality into safety and requires substantial improvement in many areas.*

0-13   *Our company safety program is very traditional and reactive with limited success potential. The application of ISO principles to safety may be just beginning.*

XXII. REFERENCES

1.   Quality Systems - model for QA in design/development, production, installation and servicing, ISO-9001:1994, International Organization for Standardization, Geneva, Switzerland, 1994.
2.   Quality management and quality system elements - guidelines, ISO-9004-1:1994, International Organization for Standardization, Geneva, Switzerland, 1994.
3.   *Accident Prevention Manual for Business and Industry, Volume 1: Administration and Program*, 10th Edition, National Safety Council, Itasca, IL, 1992.
4.   **Ferry, T.**, *Safety and Health Management Planning*, Van Nostrand Reinhold, New York, NY, 1990, 66-67.
5.   **Johnson, P. L.**, *ISO-9000: Meeting the New International Standards*, McGraw-Hill, New York, NY, 1993, 50-52.
6.   **Russell, J. P.**, *Quality Management Benchmark Assessment*, 2nd Edition, ASQC Quality Press, Milwaukee, WI, 1995, 38.
7.   Connecticut Cooperative Program, U. S. Occupational Safety and Health Administration, Draft Instruction CPL, Draft, Hartford/Bridgeport Area Office, 1996.
8.   Safety and Health Management Guidelines, U. S. Occupational Safety and Health Administration, Federal Register, 59:3904-3916, 1989.
9.   Process Safety Management, OSHA-3132, U. S. Occupational Safety and Health Administration, Washington D. C., 1994 (Reprinted).
10. **Beaumont, L. R.**, *ISO-9001, The Standard Interpretation*, 2nd Edition, ISO Easy, Middletown, NJ, 26.
11. **Kozak, R. and Krafcisin, G.**, *Safety Management and ISO-9000/QS-90000*, Quality Resources, a division of the Kraus Organization, New York, NY, 1996, 48-49.
12. **Manuele, F. A.**, Quality and safety: a reality check, *Professional Safety*, 40.6, 26, 1995.
13. **Ferry, T.**, *Safety and Health Management Planning*, 290-300.
14. **Petersen, D.**, *Analyzing Safety System Effectiveness*, 3rd Edition, Van Nostrand Reinhold, New York, NY, 1996, 158-159.
15. **Arnold, K. L.**, *The Manager's Guide to ISO-9000*, The Free Press, New York, NY, 1994, 81.
16. **Beaumont**, *ISO-9001*, 38.

17.   **Arnold**, *The Manager's Guide to ISO-9000*, 85.
18.   **Ferry**, *Safety and Health Management Planning*, 305.
19.   **Roughton, J.**, Integrating a total quality management system in safety and health programs, *Professional Safety*, 38.6, 32, 1993.
20.   **Arnold**, *The Manager's Guide to ISO-9000*, 112-113.
21.   **Arnold**, *The Manager's Guide to ISO-9000*, 121.
22.   **Arnold**, *The Manager's Guide to ISO-9000*, 123.
23.   **Arnold**, *The Manager's Guide to ISO-9000*, 157.
24.   **Johnson**, *ISO-9000*, 108-109.
25.   **Arnold**, *The Manager's Guide to ISO-9000*, 202.
26.   **Beaumont**, *ISO-9001*, 81.
27.   **Peach, R. W.**, A Quality System Checklist, in Peach, R. W., Ed. *The ISO-9000 Handbook*, 2nd Edition, Irwin Professional Publishing, Fairfax, VA, 1995.
28.   Training Requirements in OSHA Standards and Training Guidelines, OSHA 2254, U. S. Occupational Safety and Health Administration, Washington, D. C., 1995 (Revised), 4-7.
29.   Training Requirements in OSHA Standards, 8.

Chapter 4

VOLUNTARY PROTECTION PROGRAM

In this chapter the requirements of the OSHA Voluntary Protection Program (VPP) are presented and discussed. The VPP topic areas are:

- Safety Management Program
- Safety Policy and Objectives
- Management Commitment and Involvement
- Employee Participation
- Assigned Responsibilities and Accountabilities
- Contract Workers
- Program Review
- Work-site Hazards Analysis
- Baseline Surveys
- Site Inspections
- Hazard Prevention and Control
- Procedures and Protective Equipment
- Facility and Equipment Maintenance
- Hazard and Program Communications
- Accident and Injury Analysis
- Emergency Planning
- Medical Program
- Safety and Health Training
- Safety and Health Training Content
- Supervisor and Manager Readiness

In discussing the VPP guidelines, the Occupational Safety and Health Administration (OSHA) has concluded that effective management of worker safety and health protection is a decisive factor in reducing the extent and the severity of work-related injuries and illnesses. Effective management addresses all work-related hazards, including those potential hazards which could result from a change in work-site conditions or practices. It addresses hazards whether or not they are regulated by government standards.

In commenting on the guideline OSHA cites the following: *Over the years, OSHA and State enforcement and consultation staff have seen many examples of exemplary workplaces where safety and health programs were well managed and where injury rates were exceptionally low. The common characteristics observed at these sites were the use of organized and systematic methods to assign appropriate responsibility to all managers, supervisors, and employees, to inspect regularly for and control existing and potential hazards, and to orient and train all employees in the ways and means to eliminate or avoid those hazards.*

Note: The discussion and reference material for each topic are edited versions of the OSHA Safety and Health Program Management Guidelines[1] including the commentary and defined polices and procedures.[2] Also included are notes from the DOE's Revised VPP Policies and Procedures.[3]

The original guidelines and other documents should be referred to for the complete, unedited texts.

The guidance provided in this section is targeted towards general industry. Guidance for construction activities in the OSHA documents has some minor variations, particularly involving requirements for joint labor-management committees and for more frequent site inspections.

OSHA's on-site evaluation for VPP status includes: conferences with company officials and employees, review of safety documentation, plant walkthroughs, and interviews with a wide variety of plant and contractor employees. The evaluation team looks for good working conditions, employees who are aware of site hazards and hazard control programs, employees who know how to protect themselves and communicate to management regarding safety concerns, and a management team fully involved in hazard prevention and control.

During the VPP evaluation visit, OSHA:
- reviews the OSHA 200 log for completeness and accuracy - including a verification of the information using lost workday entries and other documentary information;
- compares injury and lost workday rates with the general rates for the industry;
- reviews safety and health program documentation for the means of detecting hazards and assessing potential exposures;
- reviews the hazard communication system;
- reviews the joint labor-management committees to assess the level of the committee's activities and the extent of employee involvement;
- reviews employee complaint records to determine how the complaint system is working, and if the response were timely and appropriate;
- if necessary, reviews all the chemical process systems documentation for compliance with the Process Safety Management Standard (see Chapter 5 in this text); and
- reviews manager and supervisor performance records.

Typical evaluation visit document reviews include:
- OSHA Log, first aid logs, workers' compensation reports, employee medical records;
- company policy, goal and objective statements;
- reports identifying potential health hazards, industrial hygiene and medical surveillance records;
- training programs for safety and health (including committees and OSHA recordkeeping), and training sessions attendance records;
- self-inspection and accident reports, including tracking;
- forms used for reports of safety or health problems and suggestions and tracking systems;
- records of engineering controls and the Lockout/Tagout Program;
- preventive maintenance records;
- plant safety and health rules and emergency procedures;
- PPE and Hazard Communication Programs;
- safety and health committee minutes (where applicable);
- evidence of line accountability (management evaluations, reward or penalty systems, budget accountability, disciplinary system, etc.);
- contractor program, including contractor on-site injury records;
- internal audits or evaluations of the entire safety and health program, including analysis of progress toward statistical and structural/programmatic goals;
- hazard review and analysis documentation such as process reviews and/or job safety analyses;
- a list of all hazardous chemicals on-site; and
- copies of all Process Safety Management information, if necessary.

I. SAFETY MANAGEMENT PROGRAM

A. OSHA GUIDANCE AND DISCUSSION

The general guideline is that employers are advised and encouraged to institute and maintain in their establishments a program which provides systematic policies, procedures, and practices that are adequate to recognize and protect their employees from occupational safety and health hazards. OSHA comments that *this means that the end (protection of employees from occupational safety and health hazards) determines the means. The form of the safety and health program elements and implementing actions will vary at each site according to the nature of site organization and the nature of the hazards and potential hazards at the site.*

An effective program includes provisions for the systematic identification, evaluation, and prevention or control of general workplace hazards, specific job hazards, and potential hazards which may arise from foreseeable conditions. *Provisions for identifying and preventing hazards should be systematic. If not, hazards or potential hazards will be missed and/or preventive controls will break down, and the chance of injury or illness will significantly increase.*

General workplace hazards include tripping hazards in walking areas and poor illumination. Specific job hazards relate to the specific conditions in a job, or to the inherent hazardousness of an operation required in the job. A program should look beyond specific requirements of law to address all hazards. It will seek to prevent injuries and illnesses, simple compliance is not an issue.

OSHA recognizes that although compliance with the law is an important objective, an effective program looks beyond specific requirements of law to address all hazards. It will seek to prevent injuries and illnesses, whether or not compliance is at issue. *Although standards provide important guidance on the identification and control of hazards, they are not always enough. The most successful programs look beyond government standards and legal requirements. They look for other sources of information about hazards and they use their own seasoned analytical abilities to look for and address hazards not covered by government or other standards. Their motive is to prevent injuries and illnesses and the attendant human and economic costs, whether or not compliance with the law is at issue.*

OSHA cautions that the extent to which the program is described in writing is less important then how effective it is in practice. *Relatively simple, unwritten policies, practices, and procedures are adequate to address the hazards in many smaller or less hazardous establishments. The more complex and hazardous an operation is, the more formal (written) and complex the program will probably need to be. A written program which is revised regularly clarifies policy, creates consistency and continuity in its interpretation, serves as a checkpoint whenever there is a question of priority between safety and production, and supports fair and equitable enforcement of safe work rules and practices.*

OSHA would like written programs to also include: specific work-site hazards, engineering and administrative controls, and PPE requirements.

B. DOE GUIDANCE

The U.S. Department of Energy (DOE) in adapting the Voluntary Protection Program, requires a written safety and health program. Specifically, DOE states that *all critical elements of the safety and health program, including management leadership, employee involvement, work-site analysis, hazard prevention and control, and safety and health training, must be part of the written program.*

SAFETY MANAGEMENT PROGRAM: ASSESSMENT QUESTIONS

- Is there a safety and health program with systematic policies, procedures, and practices?
- Is the program completely documented?
- Does the written program address:
  - management leadership?
  - employee involvement?
  - work-site analysis?
  - hazard prevention and control?
  - safety training?
- Does the written program identify:
  - specific work-site hazards?
  - PPE requirements?
- Are there administrative controls to ensure program elements are completed?
- Does the program use administrative limits to prevent the reaching of safety limits?
- Does the program use safety limits and requirements more conservative than those given in regulations?
- Can the workforce describe the safety and health program?
- Does the workforce embrace the safety and health program?
- Is the safety and health program effective - has performance been consistently high or improving over time?

II. SAFETY POLICY AND OBJECTIVES

A. OSHA POLICY GUIDANCE

Management must clearly state a work-site policy on safe and healthful work and working conditions, so that all personnel with responsibility at the site and personnel at other locations with responsibility for the site understand the priority of safety and health protection in relation to other organizational values. *A statement of policy is the foundation of safety and health management. It communicates the value to which safety and health protection is held in the business organization. If it is absorbed by all in the organization, it becomes the basic point of reference for all decisions affecting safety and health. It also becomes the criterion by which the adequacy of protective actions is measured.*

B. OSHA GOAL GUIDANCE

Management must also establish and communicate a clear goal for the safety and health program and objectives for meeting that goal, so that all members of the organization understand the results desired and the measures planned for achieving them. *A goal, and implementing objectives, make the safety and health policy more specific. Communicating them ensures that all in the organization understand the direction it is taking.*

C. NATIONAL SAFETY COUNCIL GUIDANCE

The National Safety Council[4] recommends that the policy statement should reflect:
- the importance that management places on the health and well-being of employees;
- management's commitment to occupational safety and health;
- the emphasis the company places on efficient operations, with a minimum of accidents and losses;

- the intention to integrate loss control in all operations, including compliance with applicable standards;
- the necessity for active leadership, direct participation, and enthusiastic support of the entire organization.

An example is given of the AT&T policy statement which says: *No job is so important and no service so urgent - that we cannot take time to perform our work safely.* Once the policy statement is formulated, management must ensure that it is communicated so that each employee becomes familiar with its content and how it applies to him or her.

D. ADDITIONAL COMMENTARY

In a proactive environment, safety goals are established as an integral part of the organization's normal goal setting process. These safety goals are then periodically reviewed and revised, hopefully to help foster continuous improvement.

Petersen[5] outlines the general content of a corporate safety policy. It should address:

- management intent,
- the scope of activities covered,
- responsibilities,
- accountability,
- safety staff assistance,
- safety committees,
- authority, and
- standards.

SAFETY POLICY AND OBJECTIVES: ASSESSMENT QUESTIONS

- Is there a clear safety and health policy?
- Are there clear safety and health goals and specific, results-oriented objectives?
- Does everyone understand the policy, goals, and objectives?
- Can everyone describe the results desired and the means for achieving them?
- Was the workforce involved in developing the safety and health policy and goals?
- Does safety and health have a priority equal to other corporate objectives?

III. MANAGEMENT COMMITMENT AND INVOLVEMENT

A. OSHA GUIDANCE

OSHA believes that management commitment and employee involvement are complementary. Management commitment provides the motivating force and the resources for organizing and controlling activities within an organization. In an effective program, management regards workers' safety and health as a fundamental value of the organization and applies its commitment to safety and health protection with as much vigor as to other organizational purposes. Employee involvement provides the means through which workers develop and/or express their own commitment to safety and health protection, for themselves and for their fellow workers.

The safety program should provide visible top management involvement in implementing the program, so that everyone will understand that management's commitment is serious. *In commenting, OSHA describes that actions speak louder than words. If top management gives high priority to safety and health protection in practice, other will see and follow. If not, a written or spoken policy of high priority for safety and health will have little credibility, and others will not follow it. Plant managers who wear required personal protective equipment in work areas, perform periodic "housekeeping" inspections, and personally track performance in safety and health protection demonstrate such involvement.*

B. NATIONAL SAFETY COUNCIL GUIDANCE

The National Safety Council[6] notes that top management provides the motivation to get the safety program started and to oversee its operations. Its responsibility is to set objectives and policy and provide the support needed by safety personnel in regard to information, facilities, tools, and equipment to conduct an effective program and establish a safe work environment.

Management can show its commitment to safety by such activities as: managing the safety and health programs, attending safety and health meetings, periodic walk-through inspections, reviewing safety records with department heads and joint employee-management meetings, investigating accidents, providing awards, and setting good examples. Management should also audit and comprehensively review the program to ensure its implementation and effectiveness. Management must also provide adequate resources for safety including staff, equipment, and promotions.

C. DOE GUIDANCE

DOE calls for the demonstration of *top-level management commitment to occupational health and safety*. Furthermore, the DOE also states that *management systems for comprehensive planning must address health and safety. Management involvement must be apparent to all employees. This should include establishing clear lines of communication with employees, setting an example of safe and healthful behavior, ensuring that all employees have a safe and healthful workplace, and being accessible to employees for health and safety concerns.*

Management commitment also extends to the provision of professional resources. For Star Programs, there should be access to, and use of, certified health and safety professionals (certified industrial hygienists, certified safety professionals, certified safety engineers, certified occupational health nurses, and/or certified occupational medical physicians). These individuals don't have to be on-site but could be located at headquarters or provided under contract.

D. COMMITMENT EXAMPLE

At one DOE site,[7] senior-level managers participate in a "Landlord Program." Managers are designated as Landlords for portions of the site and are responsible for ensuring the areas are in good repair and free from hazards. Each Landlord develops a personal checklist for monthly walkaround inspections.

MANAGEMENT COMMITMENT AND INVOLVEMENT: ASSESSMENT QUESTIONS

- Is top management fully committed to safety and actively involved in its promotion?
- Does top management signal its commitment by adhering to safety rules and behaviors?
- Is the plant manager personally involved in safety?
- Does management: chair safety meetings? provide more than adequate budget, staff, and equipment resources? lead safety improvement efforts? enforce employee and management accountabilities? and positively reward employees?
- Are safety issues regularly discussed at management operations meetings?
- Is the management safety and health representative appropriate for the site and hazards?
- Has management provided access to certified safety and health professionals?
- Is management accessible to employees for health concerns?

IV. EMPLOYEE PARTICIPATION

OSHA requires that the safety program provides for the encouragement of employee involvement in the structure and operation of the program and in decisions that affect their safety and health, so that they will commit their insight and energy to achieving the safety and health program's goals and objectives. Employees should be meaningfully involved in the site's safety and health program. *In commenting on this requirement OSHA says that this does not mean transfer of responsibility to employees. Responsibility for safety and health protection is clearly placed on the employer. However, employees intimate knowledge of the jobs they perform and the special concerns they bring to the job give them a unique perspective which can be used to make the program more effective.*

Employee participation may take any or all of a number of forms. For instance, the system for notifying management personnel about conditions that appear hazardous serves as a major means of work-site analysis to identify hazards. Such a system is, however, by itself not sufficient to provide for effective employee involvement. Forms of participation which engage employees more fully in systematic prevention include

(1) inspecting for hazards and recommending corrections or controls;

(2) analyzing jobs to locate potential hazards and develop safe work procedures;

(3) developing or revising general rules for safe work;

(4) training newly hired employees in safe work procedures and rules, and/or training their co-workers in newly revised safe work procedures;

(5) providing programs and presentations for safety meeting; and

(6) assisting in accident investigations.

Such functions can be carried out in a number of organizational contexts. Joint labor-management committees are most common. Other means include labor safety committees, safety circle teams, rotational assignment of employees to such functions, and acceptance of employee volunteers for the functions.

Employee involvement is effective only when the employer welcomes it and provides protection from any discrimination, including unofficial harassment, to the employees involved. However, inclusion of employees in one or more of the suggested activities, or in any way that fits the individual work-site and provides an employee role that has impact on decisions about safety and health protection, will strengthen the employer's overall program of safety and health protection.

Joint labor-management committees are a common and effective way of obtaining employee involvement. Such committees may have a variety of responsibilities including conducting site inspections and accident analyses. Other typical responsibilities include: safety training, complaint response, review of new equipment and procedures, etc. Committees should have access to all relevant safety and health information including: injury, illness and first aid logs; worker's compensation records; accident investigation reports; accident statistics; safety and health complaints; industrial hygiene survey and sampling results; and training records.

OSHA expects committee meetings to be held at least monthly with review and discussions covering committee inspection results; accident investigations; safety and health complaints; injury and illness logs; and analysis of apparent injury trends. There should be written meeting minutes including actions taken, recommendations made, and members in attendance. Quorums require at least half of the membership with representatives of both employees and management attending.

For joint labor-management committees there should be a reasonably equal division of labor and management members. Committee meetings should be regularly scheduled, held as scheduled, and well attended. Committee members should understand their roles and responsibilities and should receive training on their committee role.

If committee members perform site inspections, they should be trained to recognize hazards. Inspections should be regularly scheduled and conducted, covering all site areas. All inspections should include labor representatives. If committee members perform accident analyses, they also should be appropriately trained.

EMPLOYEE PARTICIPATION: ASSESSMENT QUESTIONS

- Does management encourage and see to meaningful employee involvement in safety and health activities?
- Do employees:
    - inspect for hazards?
    - help analyze jobs?
    - help develop safe work rules?
    - train new employees?
    - participate on audits or program reviews?
    - collect monitoring samples and data?
    - participate in safety meetings?
    - assist in accident investigations?
- Are employees on safety improvement teams and safety committees?
- Do all employees perceive that they can positively affect safety activities?
- Are employees knowledgeable about:
    - the safety programs?
    - employee participation activities?
    - their access to self-inspection and accident analysis information?
- Do employees have access to all pertinent safety and health information?
- Have employees influenced safety decisions?
- Are employees encouraged to stop work that represents a serious hazard or risk?
- If joint labor-management committees are used,
    - are the members trained?
    - are the committees effective in promoting employee involvement?

V. ASSIGNED RESPONSIBILITIES AND ACCOUNTABILITIES

A. OSHA GUIDANCE
1. Responsibility

OSHA requires that the safety process assign and communicate responsibility for all aspects of the safety program so that manages, supervisors, and employees in all parts of the organization know what performance is expected of them.

Responsibility needs to be assigned throughout the organization. The assignment of responsibility for safety and health protection to a single staff member, or even a small group, will leave other members feeling that someone else is taking care of safety and health problems. Everyone in an organization has some responsibility for safety and health. A clear statement of that responsibility, as it relates both to organizational goals and objectives and to the specific functions of individuals, is essential. If all persons in an organization do not know what is expected of them, they are unlikely to perform as desired.

In conjunction with assignment, it is important to provide adequate authority and resources to responsible parties, so that assigned responsibilities can be met.

The basis is that it is unreasonable to assign responsibility without providing adequate authority and resources to get the job done. For example, a person with responsibility for the safety of a piece of machinery needs the authority to shut it down and get it repaired. Needed resources may include adequately trained and equipped personnel and adequate operational and capital expenditure funds.

2. Accountability

OSHA requires that executives, managers, supervisors, and employees be held accountable for meeting their responsibilities, so that essential tasks will be performed. Accountability must reach from the Chief Executive Officer on down.

The basis is that merely stating expectations of (executives), managers, supervisors, and other employees means little if management is not serious enough to track performance, to reward it when it is competent, and to correct it when it is not. Holding everyone accountable for meeting the provision will ensure that health protection will not be neglected. To be effective, a system of accountability must be applied to everyone, from senior management to hourly employees. If some are held firmly to expected performance and others are not, the system will lose its credibility. Those held to expectations will be resentful; those allowed to neglect expectations may increase their neglect. Consequently, the chance of injury and illness will increase.

B. DOE GUIDANCE

DOE describes that *the commitment of resources should be documented and include: staffing, space, equipment, training, and promotions.*

Department heads, supervisors, foremen, and employee representatives have important leadership responsibilities within the organization. For safety and health programs:

The department heads should make sure that materials, equipment, and tools for their areas are hazard-free and that adequate controls have been provided. They make sure equipment is being used as designed and is well maintained. They know about injury/accident trends and take proper corrective action to reverse the trends. They investigate accidents and see that all rules, regulations, and procedures are enforced.

Supervisors are generally responsible for creating a safe work setting and integrating hazard recognition and control into all work activities. They monitor carefully to prevent accidents. They must be prepared to intervene in dangerous operations and take immediate corrective action. Supervisors monitor their areas on a constant basis for the human, situational, and environmental factors which could cause accidents.

Foremen and employee representatives share responsibility with upper management. They also inspect, detect, and correct. They should encourage compliance with safety and health requirements. They take corrective action where possible and report hazards to supervision. They may participate in accident investigations and represent the employees on safety committees.

DOE expects to see accountability demonstrated through evaluation of employees at all levels. There should be a functional and operational system for rewarding good performance and correcting deficient performance in place. The accountability system must be based on some type of supervisory evaluation. It may be a performance rating system, a management by objectives system with safety and health goals, or some other comparable system.

At the Westinghouse Waste Isolation Pilot Plant (WIPP) VPP site,[8] safety was one of the elements considered in the performance appraisal system. For each manager, the weight given to safety varied with the specific job assignment, increasing where safety was a vital component of performance. The actual evaluation considered performance relative to established goals.

118

C. DISCIPLINARY SYSTEM

OSHA also expects a written disciplinary system for enforcing safety and health requirements and rules. This disciplinary system should extend to management. OSHA views a disciplinary system as an indispensable piece of a whole approach to safety and health protection. OSHA comments that there should be little need to use such a corrective disciplinary system after employees have been involved in the establishment of safe work practices and safety work rules which protect themselves from the hazards of the workplace.

DOE provides an example of a three-tiered disciplinary system where one tier is for deliberate violations of the safety code, another is for routine pranks or horseplay, and the third is for minor infractions - such as failing to immediately report an injury. The penalties range in severity from immediate discharge, to the imposition of days off, down to the giving of a verbal reprimand.

In any system it is important that the application of discipline is perceived to be fair and uniform.

ASSIGNED RESPONSIBILITIES AND ACCOUNTABILITIES: ASSESSMENT QUESTIONS

- Is responsibility for all aspects of the safety and health program assigned and clearly communicated?
- Can all managers, supervisors, and employees explain their responsibilities for the program?
- Is it clear that responsibility for safety and health is spread throughout the organization?
- Is adequate authority given to make sure responsibilities are carried out?
- Are enough resources provided to ensure effective program implementation?
- Are all managers, supervisors, and employees clearly held accountable for their safety and health responsibilities?
- Do all performance appraisals include review of safety and health responsibilities and goals?
- Is there evidence that good and bad performance is addressed appropriately?
- Are there examples where accountability has not been applied?
- Is there a written and working disciplinary system?
- Are workers with safety responsibilities protected from harassment and discrimination?

VI. CONTRACT WORKERS

A. OSHA GUIDANCE

OSHA requires that written procedures control health and safety conditions for all contract workers intermingled with site employees. Contract workers involved in regular site operations must be afforded equal protection by the health and safety program. Examples are custodial workers, "nested" maintenance contractors, and temporaries. Specialty contractors are not required to be covered in the same manner, but they should be prudently selected and informed of relevant site rules and hazards that could affect them or site employees.

Safety performance should be considered in the process of selecting on-site contractors. Contractors and subcontractors should be contractually bound to maintain effective safety programs and comply with all applicable safety rules and regulations. Provision for contract oversight authority, coordination, and enforcement should be specified. There should also be contract provisions for the prompt correction and control of hazards by the site if the contractor fails to do so. Contract provisions should also detail penalties, including dismissal, for willful or repeated non-compliance by contractors or individuals.

B. ADDITIONAL COMMENTARY

In VPP organizations, contractors follow the same safety rules and regulations as do normal employees. The performance of contractors should be audited and tracked even where the contractors do self-audits.

CONTRACT WORKERS: ASSESSMENT QUESTIONS

- Are contract, temporary, and part-time workers afforded the same protection as site employees.
- Are contractors screened for health and safety programs and performance prior to hiring?
- Do contracts bind contractors to obey all applicable rules and regulations?
- Do contracts provide for:
  - oversight and enforcement?
  - the prompt correction of hazards?
  - the submission of injury data?
  - stiff penalties for repeated and willful non-compliance?
- Are contract workers fully aware of the hazards they are exposed to?
- Are temporary employees effectively supervised and monitored?
- Are injury and illness records maintained by site management for contractors whose workers work over 500 hours per quarter?

VII. PROGRAM REVIEW

A. OSHA GUIDANCE

OSHA requires a critical review of program operations at least annually to evaluate their success in meeting the goals and objectives, so that deficiencies can be identified and the program and/or the objectives can be revised when they do not meet the goal of effective safety and health protection. All elements of the safety and health program should be evaluated in the review.

A comprehensive program audit is essential periodically to evaluate the whole set of safety and health management means, methods, and processes, to ensure that they are adequate to protect against the potential hazards at the specific work-site. The audit determines whether policies and procedures are implemented as planned and whether in practice they have met the objectives set for the program. It also determines whether the objectives provide sufficient challenge to lead the organization to meet the program goal of effective safety and health protection. When either performance or the objectives themselves are found inadequate, revisions are made. Without such a comprehensive review, program flaws and their interrelationship may not be caught and corrected.

B. DOE GUIDANCE

In adapting this requirement, DOE asks for an annual, written, narrative report, including recommendations for improvements and documented timely follow-up. The review should be comprehensive, assessing the effectiveness of each element of the safety and health program.

In another element of program review, DOE also requires that trend analyses are conducted for all safety and health program data including injury and illness experience, inspections, and employee reports of problems, to help identify systematic problems that may not be noticed when only isolated incidents are considered.

Specifically, there should be critical reviews of the self-inspection program, employee hazard notification, accident analyses, employee training, enforcement of safety rules, use of PPE, routine monitoring and sampling, and review of surveillance data. The report should identify strengths and weaknesses of the safety program and have specific recommendations for improvement. The review should be conducted by a qualified staff. All actions taken to satisfy recommendations in the report should be documented.

At the WIPP VPP site,[9] DOE notes that assessments are performed by individuals or groups who are independent of the site, and who are trained in assessment techniques. Assessment results are monitored and trended. They are also scheduled to coincide with the measurement of safety goals. Assessment reports are distributed to all managers.

PROGRAM REVIEW: ASSESSMENT QUESTIONS

- Is there an annual, comprehensive safety and health program review to evaluate success and identify deficiencies?
- Is there written guidance for the review?
- Does the review cover:
  - self-inspection effectiveness?
  - accident investigation?
  - employee participation?
  - safety training?
  - hazard communication?
  - industrial hygiene?
  - enforcement of rules?
- Is top management involved in the program review?
- Is there a narrative written report with recommendations for improvement?
- Are program review results communicated to all personnel?
- Are all identified deficiencies corrected and program elements changed as needed?
- Is the program review part of a continuous improvement process?
- Is the review conducted by qualified personnel?

VIII. WORK-SITE HAZARDS ANALYSIS

A vital part of the safety program is work-site analysis involving a variety of work-site examinations, to identify not only existing hazards but also conditions and operations in which changes might occur to create hazards. Unawareness of a hazard which stems from failure to examine the work-site is a sure sign that safety and health policies and/or practices are ineffective. Effective management actively analyzes the work and work-site, to anticipate and prevent harmful occurrences. The results of hazards analyses are improved work practices and training and preventive engineering controls

This requires an active, on-going examination and analysis of work processes and working conditions. Because many hazards are by nature difficult to recognize, effective examination and analysis will approach the work and working conditions from several perspectives. The recognition of hazards which could result from changes in work practices or conditions requires especially thorough observation and thought, both from those who perform the work and those who are specially trained for that purpose.

Identification at a work-site of those safety and health hazards which are recognized in its industry is a critical foundation for safety and health protection. Successful employers actively seek the benefit of the experience of others in their industry, through trade associations, equipment manufacturers, and other sources.

Implicit in the provision for the recommended survey, reviews, and analyses is the need for employers to seek competent advice and assistance when they lack needed expertise and to use appropriate means and methods to examine and assess all existing and foreseeable hazards. Personnel who perform comprehensive baseline and update surveys, analysis of new facilities, processes, procedures, and equipment, and job hazard analyses may require greater expertise than those who conduct routine inspections.

Analysis of new facilities, processes, materials, and equipment in the course of their design and early use provides a check against the introduction of new hazards with them. Effective management ensures the conduct of such analyses during the planning phase, just before their first use, and during the early phases of their use.

Job hazard analysis is an important tool for more intensive analysis to identify hazards and potential hazards not previously recognized, and to determine protective measures. Through more careful attention to the work processes in a particular job, analysis can recognize new points at which exposure to hazards may occur or at which foreseeable changes in practice or conditions could result in new hazards.

OSHA expects that all health and safety hazards have been identified based on a complete industrial hygiene survey or process hazards review. Additional surveys and analyses are conducted whenever processes change. The complete industrial hygiene survey should be conducted by an industrial hygienist. When industrial hygiene monitoring is required for hazards, potential hazards, or identified problem areas, the sampling program should be performed by a trained staff or contract individual following sampling, testing, and analysis procedures that are consistent with national standards. There should be a written record of all results.

At one site, all new equipment and processes go through a formal design change process which is graded by the safety significance of the change. There are four levels of change review ranging from minor to significant. Reviews can involve all safety and technical disciplines, with a formal turnover package submitted to operations on completion. Similarly, maintenance activities are controlled by a maintenance work request process. Hazards are identified and work procedures are developed with input from all involved parties. Preventive maintenance procedures are verified during first use through a program of observation by the procedure writers.

WORK-SITE HAZARDS ANALYSIS: ASSESSMENT QUESTIONS

- Is there a comprehensive and active program of site, process, and conditions hazards analysis to identify potential hazards?
- Are there written procedures and guidance for the hazards analysis?
- Are all identified hazards prioritized and then effectively corrected in a timely manner?
- Is there a process to verify the operation of the hazard identification and correction system?
- Is all hazard corrective action documented?
- Is there experience to show that hazards analysis has resulted in improvements?
- Are there current hazards analyses for all work areas?
- Are the hazards analyses available to the workforce?
- Is industry experience sought to improve the hazards analysis process?
- Are safety teams and safety professionals used to ensure that all significant hazards and risks are identified?
- Do affected employees participate in the hazards analyses?

- Are employees and supervisors trained to perform hazards analyses?
- Are all new processes, facilities, materials, and equipment analyzed for potential hazards?
- Is the industrial hygiene program:
    - documented?
    - carried out by trained individuals?
    - consistent with nationally recognized procedures?
- Is there routine periodic monitoring and sampling of identified problem areas?

IX. BASELINE SURVEYS

A. OSHA GUIDANCE

OSHA requires the conduct of comprehensive baseline work-site surveys for safety and health and periodic comprehensive update surveys:

The guidance is that a comprehensive baseline survey of the work and working conditions at a site permits a systematic recording of those hazards and potential hazards which can be recognized without intensive analysis. This baseline record provides a checklist for the more frequent routine inspections. With those hazards under control, attention can be given to the intensive analysis required to recognize less obvious hazards.

Subsequent comprehensive surveys provide an opportunity to step back from the routine check on control of previously recognized hazards and look for others. With the baseline established, these subsequent reviews are one occasion for focusing more intensive analysis in areas with the highest potential for new or less obvious hazards. The frequency with which comprehensive examinations are needed depends on the complexity, hazardousness, and changeability of the work-site. Many successful work-sites conduct such reviews on an annual or biannual basis.

B. DOE GUIDANCE

DOE specifies that the periodic comprehensive surveys are to be conducted by trained and qualified safety and health professionals; that nationally recognized procedures for all sampling, testing and analysis must be used; and that written records of results are maintained.

At the WIPP VPP site,[10] the comprehensive surveys were contracted out to a consulting firm having very experienced and certified safety professionals. They covered the complete physical plant and written safety program. Enough deficiencies were identified to require a full day's shut-down to repair.

Also, based on the baseline evaluation, all the industrial hygiene information was assembled into a complete database listing all physical and chemical agents in each work area with the corresponding medical surveillance requirements and appropriate monitoring frequencies.

BASELINE SURVEYS: ASSESSMENT QUESTIONS

- Has there been a comprehensive baseline survey of the site, facilities, and processes to ensure that all safety and health hazards have been identified?
- Were the personnel performing the baseline surveys qualified to perform the comprehensive assessment?
- Were the baseline survey personnel certified safety, industrial hygiene, or medical professionals?
- Has the baseline survey and hazards inventory been regularly updated?

X. SITE INSPECTIONS

Employers are required to provide for regular site safety and health inspection, so that new or previously missed hazards and failures in hazard controls are identified.

In commenting, OSHA says that once a comprehensive examination of the workplace has been conducted and hazard controls have been established, routine site safety and health inspections are necessary to ensure that changes in conditions and activities do not create new hazards and that hazard controls remain in place and are effective. (Routine industrial hygiene monitoring and sampling which are essential components of such inspections in many workplaces are discussed in the Work-site Analysis section).

Personnel conducting these inspections also look out for new or previously unrecognized hazards, but not as thoroughly as those conducting comprehensive surveys. The frequency and scope of these "routine" inspections depend on the nature and severity of the hazards which could be present and the relative stability and complexity of work-site operations.

Both continuous and scheduled inspections should be conducted. Continuous inspections are conducted by supervisors, employees, and maintenance personnel as part of their job responsibilities. All personnel conducting inspections should be trained in hazard recognition. Continuous inspection involves noting an apparently or potentially hazardous condition and either correcting it or making a report to initiate corrective action.

Planned inspections are deliberate, thorough, and systematic by design. They permit examination of specific items or conditions. They follow established procedure and use checklists for routine items. They can be periodic, intermittent, or special.

- *Periodic inspections*, scheduled at a suitable frequency (generally monthly), can be targeted to the whole plant or specific areas. In general, the greater the accident severity potential, the more often the inspection should be scheduled. Periodic inspections can detect unsafe conditions in time to provide appropriate correction. Results over time can show degenerative trends. The personnel performing the inspections are usually familiar with the areas and can be quick to spot problems.
- *Intermittent inspections* are made at irregular intervals to focus on problem areas or departments, or on areas where changes are being made (such as new facilities and operations).
- A *general inspection* covers places not inspected periodically such as areas never visited and where people rarely get hurt. They can cover such concerns as overhead hazards or off-shift and nighttime activities. They are also appropriate after a long shutdown.

Inspections should be carefully planned and require a sound knowledge of the organization and its processes and equipment; knowledge of relevant standards and regulations; systematic inspection steps; and methods of reporting, evaluating, and using the data. There should be procedures for inspections. The procedures should provide guidance as to responsibility, frequency and schedule of inspections, use of information sources, where and what to look for, recording of findings, to whom findings are reported, and tracking of corrective actions. Hazards identified by the site inspection should be systematically corrected in a timely manner. Checklists are effective tools and help in follow-up to ensure that corrective actions have been taken.

Inspection frequency should be determined by potential loss severity, injury potential, rate at which the item can become unsafe, past history of failures, and regulatory requirements. Inspections should be conducted at least monthly with the whole site covered at least quarterly. OSHA recommends even more frequent inspections.

As discussed by Loud,[11] the inspection philosophy should not be one of compliance but one of performance. Compliance-type inspections have not resulted in significant and continuous safety improvement. They are performed because they fit in with what safety professionals know, they are easy to do, they are done by the regulators, and because they have always been done. Performance-based inspections on the other hand, focus on safety-significant activities and observations of people - and foster continuous and significant improvement in safety results

SITE INSPECTIONS: ASSESSMENT QUESTIONS

- Is there a regular, routine program of site inspections to identify workplace hazards, unsafe conditions, and improper behaviors?
- Are there written procedures or guidelines for inspections including checklists?
- Do managers, supervisors, and workers participate in the various inspections?
- Do the personnel conducting the inspections have the appropriate training, experience, and qualifications to do so?
- Are the inspections planned and overseen by a certified safety or health professional?
- Are site inspections conducted at least monthly?
- Are the inspections varied so that all site areas are effectively covered at least quarterly?
- Do the inspection reports clearly indicate conditions, necessary recommendations, and persons responsible for correction?
- Are inspection results reported to management and are hazards, nonconformances, and problems appropriately corrected?
- Is there a tracking system that ensures hazards are tracked until corrected?

XI. HAZARD PREVENTION AND CONTROL

A. OSHA GUIDANCE

OSHA describes that hazard prevention and controls are triggered by a determination that a hazard or potential hazard exists. Where feasible, hazards are prevented by effective design of the jobsite or job. Where it is not feasible to eliminate them, they are controlled to prevent unsafe and unhealthful exposure. Elimination or control is accomplished in a timely manner, once a hazard or potential hazard is recognized.

Effective management prevents or controls identified hazards and prepares to minimize the harm from job-related injuries and illnesses when they do occur.

The guideline states: so that all current and potential hazards, however detected, are corrected or controlled in a timely manner, established procedures for that purpose, using the following measures:

- engineering techniques where feasible and appropriate;
- provision of personal protective equipment; and
- administrative controls, such as reducing the duration of exposure,

should be used.

Hazards, once recognized, are promptly prevented or controlled. Management action in this respect determines the credibility of its safety and health management policy and the usefulness of it entire program.

An effective program relies on the means for prevention or control which provides the best feasible protection of employee safety and health. It regards legal requirements as a minimum. When there are alternative ways to address a hazard, effective managers have found that involving employees in discussions of methods can identify useful prevention and control measures, serve as a means for communicating the rational for decisions made, and encourage employee acceptance of the decisions.

Factors which may affect the time required for correction of hazards include:
(1) the complexity of the abatement technology;
(2) the degree of risk; and
(3) the availability of necessary equipment, materials, and staff qualified to complete the correction.

Because conditions affecting hazard correction and control vary widely, it is impractical of OSHA to recommend specific time limits for all situations. An effective program corrects hazards in the shortest time permitted by the technology required and the availability of needed personnel and materials. It also provides for interim protection when immediate correction is not possible.

B. DOE GUIDANCE

For its facilities, DOE clarifies the means and order of controlling hazards. First, use process or material substitution. Then use engineering controls. Then use administrative controls. Finally, use personal protective equipment.

At the WIPP[12] site, there is a consistent campaign to eliminate chemicals or reduce the quantities used, all managed through the hazard communication program. All MSDS data are entered into a database to allow quantities of chemicals to be tracked. All purchase requests are reviewed by industrial hygiene to see if the chemicals ordered should be allowed on site.

HAZARD PREVENTION AND CONTROL: ASSESSMENT QUESTIONS

- Is there a process of hazard prevention and control triggered when a hazard is identified?
- Are potential hazards prevented by engineering controls or by design when feasible?
- Are hazards controlled by safe work procedures when it is not feasible to prevent them?
- Is strict compliance to safety rules and standards demanded of all employees?
- Are all hazard controls in place and continually improved upon?
- Does the workforce know, understand, and support all hazard controls in effect?
- Are documented hazard reviews done by certified safety and health professionals?
- Do the hazard reviews cover:
  - machine guarding?
  - energy lockout?
  - ergonomics?
  - materials handling?
  - bloodborne pathogens?
  - confined space?
  - hazard communications?
  - other applicable standards?

XII. PROCEDURES AND PROTECTIVE EQUIPMENT

A. PROCEDURES

An important aspect of safety management is the availability of procedures for safe work which are understood and followed by all affected parties, as a result of training, positive reinforcement, correction of unsafe performance, and, if necessary, enforcement through a clearly communicated disciplinary system.

OSHA comments that when safe work procedures are the means of protection, ensuring that they are followed becomes critical. Ensuring safe work practices involves discipline. Health protection is a fundamental value of the organization. Such an environment depends on the credibility of management's involvement in safety and health matters, inclusion of employees in decisions which affect their safety and health, rigorous work-site analysis to identify hazards and potential hazards, stringent prevention and control measures, and thorough training. In such an environment, all personnel will understand the hazards to which they are exposed, why the hazards pose a threat, and how to protect themselves and others from the hazards. Training for this purpose is reinforced by encouragement of attempts to work safely and by positive recognition of safe behavior.

If, in such a context, an employee, supervisor, or manager fails to follow a safe procedure, it is advisable not only to stop the unsafe action but also to determine whether some condition of the work has made it difficult to follow the procedure or whether some management system has failed to communicate the danger of the action and the means for avoiding it. If the unsafe action was not based on an external condition or a lack of understanding, or if, after such external condition or lack of understanding has been corrected, the person repeats the action, it is essential that corrective discipline be applied. To allow an unsafe action to continue not only continues to endanger the actor and perhaps others; it also undermines the positive discipline of the entire safety and health program. To be effective, corrective discipline must be applied consistently to all, regardless of role or rank; but it must be applied.

B. PERSONAL PROTECTIVE EQUIPMENT

OSHA expects the PPE program to have strictly enforced rules that determine when to use PPE and what type to use. Depending on the hazards, eye protection, hearing protection, and breathing protection should be addressed as well as hard hats, safety shoes, and other protective equipment. The PPE program should address responsibility, availability, fit, and maintenance. Where respirators are needed, there should be a written and implemented respirator program.

PROCEDURES AND PROTECTIVE EQUIPMENT: ASSESSMENT QUESTIONS

- Are there safe work procedures for all significant job functions?
- Are there written safety rules?
- Do all personnel understand the procedures they need to use?
- Do the procedures cover the handling of any hazardous materials in the workplace?
- Are all procedures followed all the time?
- Are procedural short-cuts tolerated?
- Is there a written PPE program addressing:
  - responsibility?
  - availability?
  - fit?
  - maintenance?
- If respirators are needed, is there a written and implemented respirator program?
- Are all necessary PPE provided and used properly?

XIII. FACILITY AND EQUIPMENT MAINTENANCE

A. OSHA GUIDANCE

OSHA requires that management should provide for facility and equipment maintenance, so that hazardous breakdown is prevented.

Maintenance of equipment in facilities is an especially important means of anticipating potential hazards and preventing their development. Planning, scheduling, and tracking preventive maintenance activities provide a systematic way of ensuring that they are not neglected.

Preventive maintenance is the orderly, uniform, continuous, and scheduled action to prevent breakdown and prolong useful life of equipment and buildings. Advantages to be gained from preventive maintenance include safer working conditions, decreased equipment downtime caused by breakdown, and increased equipment life. Preventive maintenance has four main components: scheduling and performing preventive maintenance functions; keeping records of service and repairs; repairing and replacing equipment and equipment parts; and providing spare parts control.

Factors usually considered in setting up a maintenance schedule include manufacturers' recommendations, age of the machine, number of hours per day the machine is used, past experience, and machine changes with use.

Sound, efficient maintenance management anticipates machine and equipment deterioration and has procedures designed to correct defects as they develop. Such a repair system requires close integration of maintenance with inspection.

Management should keep two types of maintenance records: first, a service schedule for each piece of equipment which indicates the date purchased or put into operation, its cost, where it is used, each part to be serviced, the frequency of service, and the person assigned to do the servicing. Each piece of equipment should also have a repair record including an itemized list of parts replaced or repaired and the name of the person doing the work. The second maintenance record itemizes spare parts stock. Management should conduct routine surveys of spare parts requirements. Management should also schedule review and reordering of stocked spare parts.

Personnel assigned repair responsibilities require special safety training. Many of the jobs performed require testing or working on equipment with guards and safety devices removed. A statement of necessary precautions should be part of any repair procedure.

B. DOE OBSERVATIONS

At the WIPP site,[13] DOE observed the following maintenance practices. All plant equipment was in good working order. Equipment in the preventive maintenance program had thorough equipment history records. All preventive maintenance was on a computerized site-wide scheduling system with maintenance frequencies based on regulatory and manufacturers' requirements. A predictive maintenance program was being conducted using thermography to detect incipient failures. In addition, employees evidenced ownership of the plant's equipment through a zone system. The system assigns ownership of equipment in a defined zone to a professional or manager located in that area. No equipment was unowned.

C. NATIONAL SAFETY COUNCIL OBSERVATION

The National Safety Council[14] states that the difference between a mediocre maintenance program and a superior one is that while the mediocre one is aimed at maintaining facilities, the superior program is aimed at improving them.

FACILITY AND EQUIPMENT MAINTENANCE: ASSESSMENT QUESTIONS

- Is there a comprehensive program of facility and equipment maintenance that maximizes equipment reliability?
- Is there a preventive maintenance program for all equipment with maintenance planned, scheduled, and tracked?
- Is maintenance scheduled based on manufacturer's recommendations and actual performance?
- For important equipment, are there maintenance history records documenting surveillance, maintenance, and problems?
- Is there any sort of engineered maintenance for equipment (vibration analysis, thermography, etc.)?
- Are equipment repairs expedited for safety-related equipment?
- Do operators report maintenance needs?
- Are personnel trained to appropriately perform maintenance activities?
- Are the facility and equipment clearly well maintained?

XIV. HAZARD AND PROGRAM COMMUNICATIONS

A. OSHA GUIDANCE

In terms of communication for safety, OSHA's guidance states that: so that employee insight and experience in safety and health protection may be utilized and employee concerns may be addressed, (management should) provide a reliable system for employees, without fear of reprisal, to notify management personnel about conditions that appear hazardous and to receive timely and appropriate responses; and encourage employees to use the system.

The reasoning is that this gives management the benefit of many more points of observation and more experienced insight in recognizing hazards or other symptoms of breakdown in safety and health protection systems. It also gives employees assurance that their investment in safety and health is worthwhile.

A system is reliable only if it ensures employees a credible and timely response. The response will include both timely action to address any problems identified and a timely explanation of why particular actions were or were not taken.

Since the employer benefits from employee notices, effective management will not only guard against reprisals to avoid discouraging them but will take positive steps to encourage their submission.

The employee notification system should be written and there should be a method to track identified hazards until corrective action has been completed.

OSHA requires written assurance that employees with safety and health duties are protected from discriminatory actions including unreasonable acts of discipline, but also individual and unofficial harassment such as unpleasant or isolated duty assignments.

B. DOE OBSERVATIONS

At one site, the DOE observed five distinct employee communication mechanisms:
- a process improvement program;
- safety committee meetings;
- safety observation forms;
- a safety hotline; and
- an employee concern program.

All identified concerns were tracked in writing and usually responded to within three days. In most cases, concerns were identified to first-line supervision and cleared up without requiring a formal program for resolution.

HAZARD AND PROGRAM COMMUNICATIONS: ASSESSMENT QUESTIONS

- Are systems in place for employees to notify management of hazardous and unsafe conditions?
- Do employees use the systems without fear of harassment or discrimination?
- Are employees cited or rewarded in any way for problem notifications?
- Does management respond to employee notifications in writing? in a timely manner?
- Do employees counsel each other on safety?
- Does management use effective and innovative methods to communicate on the health and safety program?
- Do employees know the results of accident and injury investigations and corrective actions?
- Do employees know of injury/illness trend results and high hazard areas?
- Do employees know of corrective actions applied to hazards?
- Has there been any harassment of employees because of raising safety concerns?

XV. ACCIDENT AND INJURY ANALYSIS

Management should also provide for investigation of accidents and "near-miss" incidents, so that their causes and means for their prevention are identified. (OSHA requires investigation of all lost and restricted time accidents, and recommends investigation of "near-misses.")

This is needed since accidents, and incidents in which employees narrowly escape injury, clearly expose hazards. Analysis to identify their causes permits development of measures to prevent future injury or illness. Although a first look may suggest that "employee error" is a major factor, it is rarely sufficient to stop there. Even when an employee has disobeyed a required work practice, it is critical to ask, "Why?" A thorough analysis will generally reveal a number of deeper factors, which permitted or even encouraged an employee's action. Such factors may include a supervisor's allowing or pressuring the employee to take short cuts in the interest of production, or inadequate equipment, or a work practice which is difficult for the employee to carry out safely. An effective analysis will identify actions to address each of the causal factors in an accident or "near-miss" incident.

Management should also analyze injury and illness trends over time, so that patterns with common causes can be identified and prevented.

A review of injury experience over a period of time may reveal patterns of injury with common causes which can be addressed. Correlation of changes in injury experience with changes in safety and health program operations, personnel, and production processes may help to identify causes.

Accident analyses are conducted to determine direct causes, uncover indirect causes, prevent similar accidents, document facts, provide information on costs, and promote safety measures and standards. The purpose is not to find fault but to find facts. If done fairly and impartially, the analysis results will be accepted by employees. Accidents should be investigated immediately to ensure accurate details and preserve evidence. Procedures should be provided by management to guide accident investigations.

Accident investigators should examine all human, situational, and environmental causes and endeavor to generate complete information on which to take actions. A report should be issued as soon as possible, containing recommendations for preventive and corrective actions.

Eight data elements comprise the minimum amount of information that should be collected about each accident: employer characteristics, employee characteristics, narrative description of the incident, characteristics of the equipment, task being performed, time factors involved, preventive measures required, and characteristics of the injury.

Another type of accident analysis is a statistical analysis to determine accident patterns so that corrective actions can be devised. In this analysis, classifications are established for various groups of data to identify key factors causing or contributing to accidents. Cases are sorted by the classified facts to reveal the principal factors concerning the accidents - helping to pinpoint the problem areas requiring attention. Statistical analysis can also be used to compare trends or results with others.

At the WIPP site,[15] accident analyses result in "lessons learned" bulletins issued to employees. In addition, the root cause evaluation reports are available to all employees. The lessons learned bulletins must be read, and require a signature to document that fact.

At this same site, trend analyses are conducted for facility inspection findings, injuries and illnesses, employee concerns, and fire protection impairments. These items are trended monthly, quarterly, and annually with results communicated to all employees. Injury and illness reports are analyzed statistically, and management is cognizant of all analyses results. Inspection reports are combined into monthly, quarterly, and annual reports and communicated to all employees.

ACCIDENT AND INJURY ANALYSIS: ASSESSMENT QUESTIONS

- Is there a concerted effort to investigate accidents and "near-misses"?
- Do the investigations seek to determine the root causes of accidents?
- Are injury and illness results trended and analyzed and are causes identified?
- Are corrective actions effective in addressing problems?
- Have the investigators been trained in techniques of:
  - accident investigation?
  - root cause determination?
  - problem solving?
- Is an on-site safety committee member involved in each accident analysis?
- Are the results of accident and injury investigations communicated to employees?
- Are the results of accident and injury investigations at least available to the employees?
- Are other types of accident analysis (such as trend or pattern analysis) effectively conducted to determine problem areas?
- Are all injury/illness data verified by external (preferably) or internal audits?
- Do the accident analyses identify management and programmatic factors as opposed to blaming workers?
- Do employees report all accidents and injuries including early signs of injuries and accidents?

XVI. EMERGENCY PLANNING

A. OSHA GUIDANCE

Management should plan and prepare for emergencies, and conduct training and drills as needed, so that the response of all parties to emergencies will be "second nature."

Planning and training for emergencies is essential in minimizing the harmful consequences of an accident or other threat if it does occur. If personnel are not so thoroughly trained to react to emergencies that their responses are immediate and precise, they may expose themselves and others to greater danger rather than reduce their exposure. The nature of potential emergencies depends on the nature of site operations and its geographical location. The extent to which training and drills are needed depends on the severity and complexity of the emergencies which may arise.

B. NATIONAL SAFETY COUNCIL GUIDANCE

In discussing emergency planning the National Safety Council[16] states that emergency planning can minimize the potential loss from natural or human-caused disasters and accidents. Although emergency planning is often assigned to the safety professional, responsibility ultimately resides with top management.

Emergency planning must provide for the safety of employees and the public, protect property and the environment, and establish methods to restore operations to normal as soon as possible. Detailed and comprehensive planning should include natural and technological hazards that might strike a facility or even nearby.

In developing an emergency plan, there must be: identification and evaluation of potential disasters; assessment of the potential harm to people, property, and the environment; estimated warning time to mobilize the plan; determination of any changes in company operations; and consideration of any power supplies and utilities that may be required. Basic emergency management includes establishing a chain of command, alarm system, medical treatment plans, communication system, shutdown and evacuation procedures, and auxiliary power systems. Typical hazards include: fires and explosions, floods, hurricanes and tornados, earthquakes, civil strife and sabotage (including bomb threats), work accidents and rumors, wartime emergencies, hazardous and toxic materials, and weather-related emergencies.

Developing a working emergency management plan should be done in cooperation with other facilities in the area. Someone should be appointed emergency planning director or coordinator and a manual or similar document should be developed to cover the emergency steps that will be needed. The chain of command for emergency plans should be as short as possible and should be staffed with employees capable of responding well in pressure situations.

Written emergency procedures should cover emergency egress (exit routes, safe houses, assembly points), telephone numbers, responsibility for handling each type of emergency, emergency shut-down and start-up, PPE, and emergency medical services. Training should be provided on emergency responsibilities for each type of accident. Conducting unannounced drills on at least an annual basis is important.

An emergency equipment and materials checklist is good for organizing a rapid, effective response to emergencies. Equipment and materials can be on hand or on standby from nearby sources. Emergency medical services should be planned for, headed by a doctor or the health and safety professional.

EMERGENCY PLANNING: ASSESSMENT QUESTIONS

- Are all potential emergencies and likely weather conditions planned and prepared for with written procedures?
- Are all managers, supervisors, and employees trained and ready to immediately respond in emergencies?
- Is there a designated and trained site emergency response team?
- Are routine emergency drills used to rehearse emergency procedures?
- Is there at least one unannounced drill annually?

- Is equipment required for emergency support:
  - available?
  - regularly tested?
  - are personnel trained in its use?
- Does the emergency plan include routine spills and incidents?
- Do the emergency procedures include:
  - egress and assembly?
  - emergency telephone number?
  - detailed responsibilities?
  - PPE?
  - operational shut-down and start-up?
  - medical care and follow-up?
- Has the emergency plan been reviewed by:
  - a qualified safety professional?
  - the local fire department?
- Is the emergency plan reviewed annually and after each significant incident?

XVII. MEDICAL PROGRAM

A. OSHA GUIDANCE

An OSHA requirement is to establish a medical program which includes availability of first aid on site and of physician and emergency medical care nearby, so that harm will be minimized if an injury or illness does occur.

The availability of first aid and emergency medial care is essential in minimizing the harmful consequences of injuries and illnesses if they do occur. The nature of the services will depend on the seriousness of the injuries or health hazard exposures which may occur. Minimum requirements are addressed in OSHA standards.

B. NATIONAL SAFETY COUNCIL GUIDANCE

General guidance for first aid is provided by the National Safety Council.[17] First aid services on-site range from deluxe first aid kits to a well-staffed first aid and health facility. The first aid kits should be readily accessible. Without a doctor or nurse on site, a supervisory employee should be responsible for first aid supplies. Administration of first aid should initiate an accident investigation.

First aid categories are: emergency first aid and prompt attention. Prompt attention is used for treatment of minor injuries such as cuts, bruises, scratches, and burns.

A first aid program contains: properly trained and designated first aid personnel on every shift, first aid unit and supplies or kit approved by a physician, a first aid manual with procedures approved by a physician, list of reactions to chemicals via route of exposures, posted instructions for calling a physician or notifying the hospital that a person is on the way, posted method for transporting ill or injured employees, instructions for calling an ambulance or rescue squad, first aid record systems and follow-up procedures.

The first aid procedures, approved by a physician, should specify the treatment to apply to minor injuries. Health care staff should only render services covered by written standard procedures. The first aid attendant should be duly qualified and certified.

Occupational health professionals should be used to design and implement a health and surveillance monitoring program for employees exposed to occupational health hazards.

MEDICAL PROGRAM: ASSESSMENT QUESTIONS

- Is a medical program provided for?
- Are certified first aid trained employees available on site and on each shift?
- Are physician and emergency care located nearby?
- Are all supervisors and employees prepared to render or obtain aid?
- Do medical care providers participate fully in hazard identification, assessment, and training?
- Are medical personnel trained in:
  - the site's hazards analysis?
  - early recognition and treatment of illness and injuries?
  - limiting the severity of harm from an illness or injury?
- Do the medical personnel regularly review site areas?
- Does the medical program meet all site needs, including:
  - hearing conservation?
  - respirator fitting?
  - required medical testing?
- Are employee concerns about the medical program documented and responded to?
- Is the medical program adequate and appropriate for the physical and chemical hazards?
- Does the extent of medical surveillance exceed regulatory requirements?

XVIII. SAFETY AND HEALTH TRAINING

A. OSHA GUIDANCE

OSHA states that education and training are essential means of communicating practical understanding of the requirements of effective safety and health protection to all personnel. Without such understanding, managers, supervisors, and other employees will not perform their responsibilities for safety and health protection effectively.

It is not suggested that elaborate or formal training programs solely related to safety and health are always needed. Integrating consideration of safety and health protection into all organizational activities is the key to its effectiveness. Safety and health information and instruction is, therefore, often most effective when incorporated into other training about performance requirements and job practices, such as management training on performance evaluation, problem solving, or managing change; supervisors' training on the reinforcement of good work practices and the correction of poor ones; and employee training on the operation of a particular machine or the conduct of a specific task.

OSHA recommends that the employer ensure understanding of safety and health information by employees, supervisors, and managers. The act of training itself is not sufficient to endure practical comprehension. Some means of verifying comprehension is essential. Formal testing, oral questioning, observation, and other means can be useful. OSHA states that it has found that observing and interviewing employees, supervisors, and managers are the most effective measures for determining their understanding of what is expected of them in practice. Although there is no fully reliable means for ensuring understanding, effective safety and health management will apply the same diligence with respect to safety and health protection as is applied to ensuring an understanding of other operational requirements, such as time and attendance, production schedules, and job skills.

Management should ensure that all employees understand the hazards to which they may be exposed and how to prevent harm to themselves and others from exposure to these hazards. Employees should accept and follow established safety and health protections.

The commitment and cooperation of employees in preventing and controlling exposure to hazards are critical, not only for their own safety and health but for those of others as well. That commitment and cooperation depend on their understanding what hazards they may be exposed to, why the hazards pose a threat, and how they can protect themselves and others from the hazards. The means of protection which they need to understand includes not only the immediate protections from hazards in their work processes and locations, but also the management systems which commit the organization to safety and health protection and provide for employee involvement in hazard identification and prevention.

B. DOE COMMENTS

At VPP sites, training programs are exemplified by the involvement of management and the provision of sufficient funding and resources. At the WIPP site[18], the training program consisted of both formal and on-the-job training with supervisors and employees specifically trained to perform as on-the-job trainers. Everyone gets a full 12-hour general employee training. Contractors are fully included in the training program. A monthly schedule of training classes is published and distributed to all employees. The training department is responsible for issuing and maintaining formal certifications for hoisting/rigging, lockout/tagout, and permitted confined space entry.

C. NATIONAL SAFETY COUNCIL COMMENTS

The following principles of learning are applicable to safety training:[19] reinforcement, knowledge of results, practice, meaningfulness, selective learning, frequency, recall, primacy, intensity, transfer of training.

SAFETY AND HEALTH TRAINING: ASSESSMENT QUESTIONS

- Is there safety and health training for all site personnel (including contractors and temporary workers)?
- Does the training ensure all personnel understand the work hazards and their prevention?
- Does the training extend to the recognition of rules and standards violations and to recognition of programmatic deficiencies?
- Does the training stress reporting of all violations?
- Is the safety training conducted by knowledgeable personnel?
- Do employees participate in developing site-specific training?
- Do all trainees understand the safety information imparted?
- How is this ensured?
- Does training reinforce that safety and health are as important as other corporate priorities and requirements?
- Are all new employees properly trained prior to beginning duties?
- Is there a record keeping system to verify completeness of training and retraining?
- Does the record system support appropriate retraining, makeup training, and modifications to training?

XIX. SAFETY AND HEALTH TRAINING CONTENT

A. OSHA GUIDANCE

OSHA's Hazard Communication Standard specifies, for chemical hazards, an employer duty to inform employees about workplace hazards and to provide training that will enable them to avoid work-related injuries or illnesses. Other standards set forth training requirements, as summarized in OSHA Publication 2254, "Training Requirements in OSHA Standards and Training Guidelines." The rational for these standards requirements is, however, applicable in relation to all hazards.

Education and training in safety and health protection are especially critical for employees who are assuming new duties. This fact is reflected by the disproportionately high injury rates among workers newly assigned to work tasks. Although some of these injuries may be attributable to other causes, a substantial number are directly related to inadequate knowledge of job hazards and safe work practices. The Bureau of Labor Statistics reports that in 1979, 48 percent of workers injured had been on the job less than one year.

The extent of hazard information which is needed by employees will vary, but includes at least: (1) The general hazards and safety rules of the work-site; (2) specific hazards, safety rules, and practices related to particular work assignments; and (3) the employee's role in emergency situations. Such information and training is particularly relevant to hazards that may not be readily apparent to, within the ordinary experience and knowledge of, the employee.

B. NATIONAL SAFETY COUNCIL GUIDANCE

The National Safety Council[20] outlines a new employee orientation program which includes: policy statements, benefit packages, organized labor agreements (if applicable), safety and health policy statements, housekeeping standards, communications about hazards, personal protective equipment, emergency response procedures (fire, spill, etc.), accident reporting procedures, near-miss accident reporting, accident investigations, lockout/tagout procedures, machine guarding, electrical safety awareness, ladder use and storage (if applicable), confined space entry (if applicable) medical facility support, first aid and CPR, hand tool safety, ergonomic principles, eye wash and shower locations, fire prevention and protection, access to exposure and medical records.

C. ADDITIONAL GUIDANCE

Ferry[21] provides a summary of topics that a complete safety and health training program should contain. These topics are: concepts and philosophies of environmental health and of accident prevention; safety and health policies; plant processes; oral and written communication skills; job safety analysis and job instruction; accident investigation; relationship of safety and health to company objectives; accident, illness and injury reporting; legislative and reporting requirements; personnel selection and appraisal; job placement and training activities; hazard detection; operating procedures and rules; gathering and analyzing safety and health information; use of personal protective equipment; emergency medical treatment; and emergency planning and procedures.

SAFETY AND HEALTH TRAINING CONTENT: ASSESSMENT QUESTIONS

- Does the content of the safety and health training program include:
  - general work hazards?
  - work-specific hazards?
  - safety rules and practices?
  - the employee's emergency responsibilities?
- Does the content include all OSHA rights and access to information?
- Does training include OSHA applicable standards?
- Is training effectiveness verified for all employees?

XX. SUPERVISOR AND MANAGER READINESS

A. SUPERVISORS

So that supervisors will carry out their safety and health responsibilities effectively, (OSHA requires) that management ensures that they understand those responsibilities and the reasons for them, including:

(A) analyzing the work under their supervision to identify unrecognized potential hazards;

(B) maintaining physical protections in their work areas; and

(C) reinforcing employee training on the nature of potential hazards in their work and on needed protective measures, through continual performance feedback and, if necessary, through enforcement of safe work practices.

First-line supervisors have an especially critical role in safety and health protection because of their immediate responsibility for workers and for the work being performed. Effective training of supervisors will address their safety and health management responsibilities as well as information on hazards, hazard prevention, and response to emergencies. Although they may have other safety and health responsibilities, those listed in these guidelines merit particular attention.

B. MANAGERS

Management must also ensure that managers understand their safety and health responsibilities described under "Management Commitment and Employee Involvement," so that the managers will effectively carry out those responsibilities. *The reasoning is that because there is a tendency in some businesses to consider safety and health a staff function and to neglect the training of managers in safety and health responsibilities, the importance of managerial training is noted separately. Managers who understand both the way and the extent to which effective safety and health protection impact on the overall effectiveness of the business itself are far more likely to ensure that the necessary safety and health management systems operate as needed.*

C. NATIONAL SAFETY COUNCIL GUIDANCE

The National Safety Council's "Supervisors Development Program" outlines the content of a supervisor training program.[22] It includes: loss control for supervisors, communications, human relations, employee's involvement in safety, safety training, industrial hygiene and noise control, accident investigation, safety inspections, personal protective equipment, materials handling and storage, machine safeguarding, hand tools and portable power tools, electrical safety, and fire safety.

D. ADDITIONAL GUIDANCE

Ferry[23] provides a concise description of typical manager and supervisor safety responsibilities for which readiness must be ensured. For line management, the typical responsibilities are: communicating objectives and goals; working with staff to develop goals, rules, emergency plans, processes, and designs; developing strategies for compliance with rules and regulations; developing job procedures with supervisors; interpreting information for preventive (and corrective) application; implementing procedures for use of personal protective equipment; providing feedback to senior mangement and to staff on performance, and for reviews of accidents, illnesses, and injuries; investigating, evaluating, and reporting on accidents, injuries, and illnesses; taking appropriate corrective action on unsafe conditions; evaluating emergency plans; and coordinating training needs and ensuring that personnel are made available for training.

Supervisors understand the reasons for safety and health measures; understand, adopt, and ensure compliance with safety practices; implement regulatory requirements; develop employee safety awareness; discuss and ensure the use of safe work practices; inspect to identify unsafe conditions and practices; arrange for worker training; inform management of unsafe practices and conditions requiring management attention; and arrange for drills of emergency plans. They also discuss safety goals, involve employees in all aspects of safety, and recognize safe work behavior.[24]

SUPERVISOR AND MANAGER READINESS: ASSESSMENT QUESTIONS

- Have all managers and supervisors been formally trained on the safety and health program?
- Do all the supervisors clearly understand their safety and health responsibilities and do they carry them out?
- Do these responsibilities include:
  - work analysis to identify hazards?
  - physical protection maintenance?
  - enforcement of discipline?
  - reinforcement of employee training?
- Do supervisors coach employees on safety practices and behaviors?
- Do all managers understand and carry out their safety and health responsibilities? can they participate in the safety and health program?
- Do all supervisors and managers attend training in all subjects provided to employees under their direction?
- Do supervisors and managers participate in site inspections, evaluations, and audits?

XXI. SUMMARY - VOLUNTARY PROTECTION PROGRAM EVALUATION

Voluntary Protection Program Topic	Rating
1. Safety Management Program	
2. Safety Policy and Objectives	
3. Management Commitment and Involvement	
4. Employee Participation	
5. Assigned Responsibilities	
6. Accountability	
7. Program Review	
8. Work-site Hazards Analysis	
9. Baseline Surveys	
10. Site Inspections	
11. Hazard Prevention and Control	
12. Procedures and Protective Equipment	
13. Facility and Equipment Maintenance	
14. Hazard Communications	
15. Accident and Injury Analysis	
16. Emergency Planning	
17. Medical Program	
18. Safety and Health Training	
19. Safety and Health Training Content	
20. Supervisor and Manager Readiness	
Total Safety Program Score	

A. VOLUNTARY PROTECTION PROGRAM RATING GUIDE

(Rate each topic 0 to 4. Consider the scope and breadth of deployment and the results achieved.)

4= Basically all of our people or activities meet the stated criteria or their intent with excellent results achieved.
3= Most of our people/activities meet the criteria or their intent with good results.
2= About half of our people/activities meet the criteria or their intent with positive results.
1= Some or a few of our people/activities meet the criteria or their intent with a few positive results.
0= None of our people/activities meet the criteria and no positive results are evident.

B. VOLUNTARY PROTECTION PROGRAM SCORING

Total Score   *Safety System Evaluation*

68-80   *The VPP guidelines are completely built into our work safety system. There are few or no gaps or problems evident.*
55-67   *Our company has very progressively built the VPP guidelines into safety with sound approaches and few gaps in deployment but some fine tuning is possible in several areas.*
41-54   *Our company is progressively building VPP guidelines into safety but gaps in deployment exist and refinements are still needed.*
27-40   *Our company is somewhat progressive in applying the VPP guidelines but significant gaps in deployment exist. Improvement is still needed throughout.*
14-26   *Our company is in the early stages of building the VPP guidelines into safety and requires substantial improvement in many areas.*
0-13   *Our company safety program is very traditional and reactive with limited success potential. The application of the VPP guidelines to safety may be just beginning.*

XXII. REFERENCES

1.   Safety and Health Management Guidelines, U. S. Occupational Safety and Health Administration, Federal Register, 59:3904-3916, Washington, D. C., 1989.
2.   Revised Voluntary Protection Programs (VPP) Policies and Procedures Manual, Instruction TED 8.1a, U. S. Occupational Safety and Health Administration, Washington, D. C., 1996.
3.   Voluntary Protection Program: Part 1: Program Elements, DOE/EH-0433, U. S. Department of Energy, Washington, D. C., 1994.
4.   *Accident Prevention Manual for Business and Industry, Volume 1: Administration and Program*, 10th Ed., National Safety Council, Itasca, IL, 1992, 74.
5.   **Petersen, D.**, *Safety Management: A Human Approach*, Aloray, Fairview, NJ, 1975, 56.
6.   *Accident Prevention Manual*, 74-75.
7.   WIPP (Westinghouse Waste Isolation Pilot Plant) Report, DOE VPP On-site Review - August 29 - September 2, 1994, U. S. Department of Energy, Washington, D. C., September 19, 1994.

140

8.   WIPP Report, DOE VPP On-site Review.
9.   WIPP Report, DOE VPP On-site Review.
10.   WIPP Report, DOE VPP On-site Review.
11.   **Loud, J.**, Compliance inspections: unhealthy for your career?, *Professional Safety*, 41.8, 35, 1996.
12.   WIPP Report, DOE VPP On-site Review.
13.   WIPP Report, DOE VPP On-site Review.
14.   *Accident Prevention Manual*, 78.
15.   WIPP Report, DOE VPP On-site Review.
16.   *Accident Prevention Manual*, Ch. 6.
17.   *Accident Prevention Manual*, Ch. 4.
18.   WIPP Report, DOE VPP On-site Review.
19.   *Accident Prevention Manual*, 360-362.
20.   *Accident Prevention Manual*, 369.
21.   **Ferry, T.**, *Safety and Health Management Planning*, Van Nostrand Reinhold, New York, NY, 1990, 115.
22.   *Accident Prevention Manual*, 376.
23.   **Ferry**, *Safety and Health Management Planning*, 116.
24.   **Petersen, D.**, *Analyzing Safety System Effectiveness*, Van Nostrand Reinhold, New York, NY, 1996, 92.

Chapter 5

PROCESS SAFETY MANAGEMENT GUIDELINES

In this chapter the requirements of OSHA's process safety management (PSM) guidelines are presented. The process safety management areas are:

- General Guidelines
- Employee Involvement
- Process Safety Information
- Process Hazards Analysis
- Operating Procedures and Practices
- Employee Training
- Contractors
- Pre-Startup Safety
- Mechanical Integrity
- Nonroutine Work Authorizations
- Managing Change
- Investigation of Incidents
- Emergency Preparedness
- Compliance Audits

Although the process safety management guidance presented by OSHA and the EPA is specifically directed to processes involving hazardous chemicals posing a threat to workers and the public, the ideas are general and apply to all processes that can affect worker safety. For example, when making process changes, the changes should be carefully reviewed beforehand for safety implications, and when made, operators should be trained to properly use the new procedures and new equipment. In practice, the extent of application of these guidelines to other than high hazard processes should be based on the types of hazards present and the degree of risk posed by the processes.

As pointed out in the OSHA compliance guidelines,[1] auditing of the process safety management guidelines may be enhanced by including an evaluation of an activity involving the interrelationship of all or most of the program elements. For example, evaluating how a piece of process equipment has been replaced by a new, nonidentical item could include: **managing change**, new **operating procedures** with required **employee training**, and material made available for **employee** review **(involvement)**. The act of installing the equipment may involve **contractors** and **nonroutine work authorizations**. Starting the process up will require pre-startup safety. The **emergency plan** may need review for updating. The **process safety information** will require updating and the **process hazards analysis** needs verification. The new equipment must be accounted for in the **mechanical integrity** program.

Note: Included with the questions in each topic area is a *slightly edited* version of the OSHA process safety management guidelines[2,3] including comments and guidance. OSHA explanatory material is indicated by italics. Please refer to the original guidelines for the complete, unedited text.

141

OSHA's on-site PSM evaluations include: an opening conference with safety/process officials, an overview of the PSM program, an initial walkaround to gain familiarity with the site and its hazards, a review of PSM documentation, interviews with plant and contractor employees, selection of a process to evaluate against the standard, and additional walkarounds to assess compliance of the process selected against the standard.

OSHA's site evaluations are conducted by trained compliance safety and health officers (CSHOs). PSM inspection team leaders must have experience in the chemical (or explosives) industry plus specific OSHA and advanced training courses. PSM team members must be experienced inspectors, and must have completed specific OSHA courses. Other, less experienced, CSHOs may conduct programmatic PSM inspections.

Comprehensive inspections use a three-fold, program, quality, and verification (PQV) approach. The employer's **program** for PSM compliance is evaluated. The **quality** of the procedures is assessed compared to industry practices. Then effective implementation of the program is **verified** through reviews of the written program and activity records, interviews, and observations.

Typical documentation reviews for evaluation visits include:

- OSHA 200 logs for the employer and process-related contractor employees.
- The employer's written plan for employee participation.
- Written process safety information such as flow diagrams, piping and instrumentation diagrams (P&IDs), and process descriptions.
- Process Hazards Analysis (PHA) information: the rationale and order for conducting PHAs, any completed PHAs, the names of PHA team members, actions and schedules to promptly address PHA findings, resolution of findings, verification of communications to appropriate personnel, and the 5-year revalidation of the PHA.
- Procedures for operations in selected units, annual procedure certifications, and safe work procedures for hazardous operations (lockout/tagout, confined space entry, opening process lines, conducting lifts over process lines, etc.).
- Training records including: initial and refresher training records for process operators; methods for determining training content; methods for determining refresher frequency; certification of required knowledge, skills and ability to perform job of grandfathered employees; and training material.
- Pre-startup safety review and training for new and significantly modified facilities.
- Process equipment integrity materials including: procedures and schedules; sections of manufacturer's instructions, codes, and standards; and selected inspection and test records.
- Hot work permit program and active permits for the process selected.
- Written management of change materials.
- Incident investigation reports for the selected process including corrective actions.
- The emergency plan.
- The most recent audit reports with responses and verification of corrective actions.
- Contractor-related information including: contractor's safety performance and programs; methods of informing contract employers of known hazards and the emergency action plan; safe work practices to control the entrance, presence, and exit of contract employees; evaluation of contract employer performance with respect to the PSM standard; contract employee injury and illness logs for process-related work; and a list of unique hazards related to the contractor's work.
- Contractor records including: contract employee training records, known job hazards and applicable provisions of the emergency plan, and descriptions of unique hazards in the contractor's work area.
- The selection of one or more process for evaluation considering: walkthrough observations, incident reports and other process history, the presence of completed PHAs, age of process unit, nature and quantity of process chemicals, employee representative input, current hot work, equipment replacement, or other activities, and the number of employees.

I. GENERAL PROCESS SAFETY GUIDELINES

OSHA's process safety management standard contains requirements for preventing or minimizing the consequences of catastrophic releases of toxic, reactive, flammable, or explosive chemicals. These releases may result in toxic, fire or explosion hazards.

A. APPLICATION
(1) *(The OSHA process safety management standard)* applies to the following:

(i) A process which involves a chemical at or above . . . threshold quantities listed . . . .

(ii) A process which involves a flammable liquid or gas on-site in one location, in a quantity of 10,000 pounds or more except for:

(A) Hydrocarbon fuels used solely for workplace consumption as a fuel . . . if . . . not a part of a process containing another highly hazardous chemical . . . . ;

(B) Flammable liquids stored in atmospheric tanks or transferred which are kept below their normal boiling point without . . . chilling or refrigeration.

B. PURPOSE OF GUIDELINES
OSHA's introduction to Process Safety Management states that: The major objective of process safety management . . . is to prevent unwanted releases of hazardous chemicals especially into locations which could expose employees and others to serious hazards. An effective process safety management program requires a systematic approach to evaluating the whole process. Using this approach the process design, process technology, operational and maintenance activities and procedures, nonroutine activities and procedures, emergency preparedness plans and procedures, training programs, and other elements which impact the process are all considered in the evaluation. The various lines of defense that have been incorporated into the design and operation of the process to prevent or mitigate the release of hazardous chemicals need to be evaluated and strengthened to assure their effectiveness at each level. Process safety management is the proactive identification, evaluation and mitigation or prevention of chemical releases that could occur as a result of failures in process, procedures or equipment.

The process safety management standard targets highly hazardous chemicals that have the potential to cause a catastrophic incident. This standard as a whole is to aid employers in their efforts to prevent or mitigate episodic chemical releases that could lead to a catastrophe in the workplace and possibly to the surrounding community. To control these types of hazards, employers need to develop the necessary expertise, experiences, judgement, and proactive initiative within their workforce to properly implement and maintain an effective process safety management program as envisioned in the OSHA standard. This OSHA standard is required by the Clean Air Act Amendments as is the Environmental Protection Agency's Risk Management Plan. Employers, who merge the two sets of requirements into their process safety management program, will better assure full compliance with each as well as enhancing their relationship with the local community.

While OSHA believes process safety management will have a positive effect on the safety of employees . . . smaller businesses which may have limited resources available to them . . . might consider alternative avenues of decreasing the risks (of) highly hazardous chemicals at their workplaces. One method . . . is the reduction in the inventory of the highly hazardous chemical. This reduction in inventory will result in a reduction of the risk or potential for a catastrophic incident. Also, employers, including small employers, may be able to establish more efficient inventory control by reducing the quantities of highly hazardous chemicals on-site below the established threshold quantities. This reduction can be accomplished by ordering smaller shipments and maintaining the minimum inventory necessary for efficient and safe operation. When reduced inventory is not feasible, then the employer might consider dispersing inventory to several locations on-site. Dispersing storage into locations where a release in one location will not cause a release in another location is a practical method to also reduce the risk or potential for catastrophic incidents.

GENERAL PROCESS SAFETY GUIDELINES: ASSESSMENT QUESTIONS

- Is there a process safety management program for the proactive identification, evaluation and mitigation or prevention of process, procedure, or equipment accidents?
- Does the process evaluation include all activities involving the process?
- Does process evaluation include all lines of process defense?
- Does the workforce have the experience, expertise, and judgment to implement and maintain the process safety management program?
- Have alternative methods of reducing process risk been considered?

II. EMPLOYEE INVOLVEMENT

OSHA cites that the intent of this section is to require employers to involve employees at an elemental level of the PSM program.

A. OSHA REQUIREMENTS

The OSHA standard requires that:

(1) Employers shall develop a written plan of action regarding the implementation of the employee participation required . . . .

(2) Employers shall consult with employees and their representatives on the conduct and development of process hazards analyses and on the development of the other elements of process safety management . . . .

Consultation is considered by OSHA to be a two-way dialog between the employer and the employee in which the employer elicits and responds to employee's concerns and suggestions bearing upon the elements of process safety management. It is a process of seeking advice, criticisms, and suggestions from the employees.

Also, OSHA wants employees of contractors to be consulted to the same extent as are similarly situated direct hire employees. Thus the employer must establish a method for informing all contractor employees that their process safety concerns and suggestions are welcome, and will be responded to.

(3) Employers shall provide to employees and their representatives access to process hazard analyses and to all other information . . . developed . . . .

The intent of this provision is for the information to be provided in a reasonable manner. Reasonable access could require providing copies or loaning documents. Also, OSHA considers that equal access must be provided to employees of covered contractors. Contract employers share responsibility for assuring that their employees are provided with the requested information.

B. BACKGROUND

OSHA's background on employee involvement in Process Safety Management, OSHA states that Section 304 of the Clean Air Act Amendments requires employers to consult with their employees and their representatives regarding the employers' efforts in the development and implementation of the process safety management program elements and hazard assessments. Section 304 also requires employers to train and educate their employees and to inform affected employees of the findings from incident investigations required by the process safety management program. Many employers, under their safety and health programs, have already established means and methods to keep employees and their representatives informed about relevant safety and health issues and employers may be able to adapt these practices and procedures to meet their obligations under this standard. Employers who have not implemented an occupational safety and health program may wish to form a safety and health committee of employees and management representatives to help the employer meet the obligations specified by his standard. These committees can become a significant ally in helping the employer to implement and maintain an effective process safety management program for all employees.

For compliance, OSHA requires interviews to show that employees have been consulted on the process hazards analysis and the process safety management program. Employees should also have been informed of their right of access to the process hazards analysis and all the information developed regarding process safety management. OSHA also requires no unreasonable delays in providing access to the information.

EMPLOYEE INVOLVEMENT: ASSESSMENT QUESTIONS

- Is there a written employee participation plan?
- Have employees been consulted on the development of process hazards analyses and elements of process safety management?
- Do all employees (including contractor employees) have access to process hazards analyses and other process safety information?
- Are all employees appropriately trained to participate and understand the information?
- Are all affected employees informed about incident investigation findings?
- Is there a safety and health committee with employees and management participating to implement the process safety management program?

III. PROCESS SAFETY INFORMATION

The OSHA standard requires that. . . the employer shall complete a compilation of written process safety information before conducting any process hazard analysis . . . . The compilation of written process safety information is to enable the employer and the employees involved in operating the process to identify and understand the hazards posed by those processes involving highly hazardous chemicals. This process safety information shall include information pertaining to the hazards of the highly hazardous chemicals used or produced by the process, information pertaining to the technology of the process, and information pertaining to the equipment in the process.

A. OSHA REQUIREMENTS
(1) Information pertaining to the hazards of the highly hazardous chemicals in the process . . . shall consist of at least the following:
- (i) Toxicity information;
- (ii) Permissible exposure limits;
- (iii) Physical data;
- (iv) Reactivity data;
- (v) Corrosivity data;
- (vi) Thermal and chemical stability data; and
- (vii) Hazardous effects of inadvertent mixing of different materials that could foreseeably occur.

Note: Material Safety Data Sheets . . . may be used to comply with this requirement to the extent they contain the information required . . . .

(2) Information pertaining to the technology of the process.

(i) Information concerning the technology of the process shall include . . . .

(A) A block flow diagram or simplified process flow diagram . . .;

(B) Process chemistry;

(C) Maximum intended inventory;

(D) Safe upper and lower limits for such items as temperatures, pressures, flows, or compositions; and,

(E) An evaluation of the consequences of deviations, including those affecting the safety and health of employees.

(ii) Where the original technical information no longer exists, such information may be developed in conjunction with the process hazard analysis in sufficient detail to support the analysis.

(3) Information pertaining to the equipment in the process.

(i) Information pertaining to the equipment in the process shall include:

(A) Materials of construction;

(B) Piping and instrument diagrams (P&ID's);

(C) Electrical classification;

(D) Relief system design and design basis;

(E) Ventilation system design;

(F) Design codes and standards employed;

(G) Material and energy balances . . . ; and,

(H) Safety systems (e.g., interlocks, detection or suppression systems).

(ii) The employer shall document that equipment complies with recognized and generally accepted good engineering practices.

(iii) For existing equipment designed and constructed in accordance with codes, standards, or practices . . . no longer in general use, the employer shall determine and document that the equipment is designed, maintained, inspected, tested, and operating in a safe manner.

OSHA wants the process safety information retained for the lifetime of the process and updated whenever changes other than replacement in kind are made. If data on the original technology is not available - the employer must obtain or generate the missing information in time to perform the PHA.

B. BACKGROUND

1. Purpose of Information

OSHA's background on process safety information states that complete and accurate written information concerning process chemicals, process technology, and process equipment is essential to an effective process safety management program and to a process hazards analysis. The compiled information will be a necessary resource to a variety of users including the team that will perform the process hazards analysis as required . . . ; those developing the training programs and the operating procedures; contractors whose employees will be working with the process; those conducting the pre-startup reviews; local emergency preparedness planners; and insurance and enforcement officials.

The information to be compiled about the chemicals, including process intermediates, needs to be comprehensive enough for an accurate assessment of the fire and explosion characteristics, reactivity hazards, the safety and health hazards to workers, and the corrosion and erosion effects on the process equipment and monitoring tools. Current material safety data sheet (MSDS) information can be used to help meet this requirement which must be supplemented with process chemistry information including runaway reaction and over pressure hazards if applicable.

2. Process Diagrams

Process technology information will be a part of the process safety information package and it is expected that it will include diagrams . . . as well as employer established criteria for maximum inventory levels for process chemicals; limits beyond which would be considered upset conditions; and a qualitative estimate of the consequences or results of deviation that could occur if operating beyond the established process limits. Employers are encouraged to use diagrams which will help users understand the process.

A block flow diagram is used to show the major process equipment and interconnecting process flow lines and show flow rates, stream composition, temperatures, and pressures when necessary for clarity. The block flow diagram is a simplified diagram.

Process flow diagrams are more complex and will show all main flow streams including valves to enhance the understanding of the process, as well as pressures and temperatures on all feed and product lines within all major vessels, in and out of headers and heat exchangers, and points of pressure and temperature control. Also, materials of construction information, pump capacities and pressure heads, compressor horsepower and vessel design pressures and temperatures are shown when necessary for clarity. In addition, major components of control loops are usually shown along with key utilities on process flow diagrams.

Piping and instrument diagrams (P&IDs) may be the more appropriate type of diagrams to show some of the above details and to display the information for the piping designer and engineering staff. The P&IDs are to be used to describe the relationships between equipment and instrumentation as well as other relevant information that will enhance clarity. Computer software programs which do P&IDs or other diagrams useful to the information package, may be used to help meet this requirement.

3. Design Information

The information pertaining to process equipment design must be documented. In other words, what were the codes and standards relied on to establish good engineering practice. These codes and standards are published by such organizations as the American Society of Mechanical Engineers, American Petroleum Institute, . . . and model building code groups. In addition, various engineering societies issue technical reports which impact process design. . . . . This type of technically recognized report would constitute good engineering practice.

For existing equipment designed and constructed many years ago in accordance with the codes and standards available at that time and no longer in general use today, the employer must document which codes and standards were used and that the design and construction along with the testing, inspection and operation are still suitable for the intended use. Where the process technology requires a design which departs from the applicable codes and standards, the employer must document that the design and construction are suitable for the intended purpose.

For compliance, OSHA requires that documentation show that compliance with appropriate consensus standards has been researched. For equipment designed to older codes and standards, documentation of continued acceptability may be through documentation of successful prior operation procedures, documentation that equipment is consistent with appropriate editions of codes, or performing an engineering analysis.

PROCESS SAFETY INFORMATION: ASSESSMENT QUESTIONS

- Is there complete and accurate process information addressing process:
  - hazards?
  - design bases?
  - technologies?
  - materials?
  - equipment?
- Is this information available to all who need it?
- Are there process flow diagrams or P&IDs?
- Are all codes and standards (and appropriate engineering practices) used for design documented?
- For highly hazardous chemicals, is there sufficient information regarding the hazards?

IV. PROCESS HAZARDS ANALYSIS

A. OSHA REQUIREMENTS

The OSHA standard requires that:

(1) The employer shall perform an initial process hazard analysis (hazard evaluation) on processes . . . . The process hazard analysis shall be appropriate to the complexity of the process and shall identify, evaluate, and control the hazards involved in the process. Employers shall determine and document the priority order for conducting process hazard analyses based on . . such considerations as extent of the process hazards, number of potentially affected employees, age of the process, and operating history of the process. The process hazard analysis shall be conducted as soon as possible . . . .

These process hazard analyses shall be updated and revalidated . . . .

The PHA is intended to control process hazards through the timely resolution of PHA findings and recommendations.

(2) The employer shall use one or more of the following methodologies that are appropriate to determine and evaluate the hazards of the process being analyzed.

- (i) What-If;
- (ii) Checklist;
- (iii) What-If/Checklist;
- (iv) Hazard and Operability Study (HAZOP);
- (v) Failure Mode and Effects Analysis (FMEA);
- (vi) Fault Tree Analysis; or
- (vii) An appropriate equivalent methodology.

OSHA expects that the methodology be selected based on sound judgement. It should have the capability to elicit all hazards, defects, and failure possibilities for the process being analyzed, and also have the capability to address all the factors which are required below.

(3) The process hazard analysis shall address:

(i) The hazards of the process;

Typical hazards are: toxic release, fire, explosion, runaway reaction, polymerization, overpressurization, corrosion, overfilling, contamination, equipment failure, loss of cooling, heating, electricity, instrument air, earthquake, flood, tornado, hurricane, etc. (see EPA).

(ii)   The identification of any previous incident which had a likely potential for catastrophic consequences in the workplace;

(iii)   Engineering and administrative controls applicable to the hazards and their interrelationships such as appropriate application of detection methodologies to provide early warning of releases. (Acceptable detection methods might include process monitoring and control instrumentation with alarms, and detection hardware such as hydrocarbon sensors.);

(iv)   Consequences of failure of engineering and administrative controls;

Process controls include: vents, relief valves, check valves, scrubbers, flares, manual shutoffs, automatic shutoffs, interlocks, alarms and procedures, keyed bypass, emergency air supply, emergency power, backup pump, grounding equipment, inhibitor addition, rupture disks, excess flow device, quench system, purge system, etc. *Mitigation systems include: sprinkler systems, dikes, fire walls, blast walls, deluge systems, water curtains, enclosures, and neutralization.*

(v)   Facility siting;

Where siting refers to the location of various components within the establishment.

(vi)   Human factors; and

(vii)   A qualitative evaluation of a range of the possible safety and health effects of failure of controls on employees in the workplace.

The intent is to require the employer at least to identify each type of control as well as identify the possible effects of the failure of the listed control. The employers can determine the consequences of a failure of these controls, and establish a reasonable estimate of the health and safety effects on employees without conducting a specialized quantitative evaluation.

(4) The process hazard analysis shall be performed by a team with expertise in engineering and process operations, and the team shall include at least one employee who has experience and knowledge specific to the process being evaluated. Also, one member of the team must be knowledgeable in the specific process hazard analysis methodology being used.

(5) The employer shall establish a system to promptly address the team's findings and recommendations; assure that the recommendations are resolved in a timely manner and that the resolution is documented; document what actions are to be taken; complete actions as soon as possible; develop a written schedule of when these actions are to be completed; communicate the actions to operating, maintenance, and other employees whose work assignments are in the process and who may be affected by the recommendations or actions.

Timeliness refers to all due speed, considering the complexity of the recommendation and the difficulty of implementation. Employers should develop a schedule to complete corrective actions, to document what actions are to be taken, and to document the completion of those actions as they occur. OSHA cautions that the employer can justifiably decline to adopt a recommendation where the employer can document, in writing and based upon adequate evidence, that one or more of the following conditions is true:

● *The analysis on which the recommendation is based contains material factual errors.*

● *The recommendation is not necessary to protect the health and safety of the employer's own employees, or the employees of contractors.*

● *An alternate measure would provide a sufficient level of protection.*

● *The recommendation is infeasible.*

(6) At least every five (5) years after the completion of the initial process hazard analysis, the process hazard analysis shall be updated and revalidated by a *(qualified)* team . . . , to assure that the process hazard analysis is consistent with the current process.

(7) Employers shall retain process hazards analyses and updates or revalidations for each

process . . . , as well as the documented resolution of recommendations . . . for the life of the process.

B. BACKGROUND

1. PHA Purpose

OSHA's background on process hazard analysis states that: A process hazard analysis (PHA), sometimes called a process hazard evaluation, is one of the most important elements of the process safety management program. A PHA is an organized and systematic effort to identify and analyze the significance of potential hazards associated with the processing or handling of highly hazardous chemicals. A PHA provides information which will assist employers and employees in making decisions for improving safety and reducing the consequences of unwanted or unplanned releases of hazardous chemicals. A PHA is directed toward analyzing potential causes and consequences of fires, explosions, releases of toxic or flammable chemicals, and major spills of hazardous chemicals. The PHA focuses on equipment, instrumentation, utilities, human actions (routine and nonroutine), and external factors that might impact the process. These considerations assist in determining the hazards and potential failure points or failure modes in a process.

2. PHA Selection

The selection of a PHA methodology or technique will be influenced by many factors including the amount of existing knowledge about the process. Is it a process that has been operating for a long period of time with little or no innovation and extensive experience has been generated with its use? Or, is it a new process or one which has been changed frequently by the inclusion of innovative features? Also, the size and complexity of the process will influence the decision as to the appropriate PHA methodology to use. All PHA methodologies are subject to certain limitations. For example, the checklist methodology works well when the process is very stable and no changes are made, but it is not as effective when the process has undergone extensive change. The checklist may miss the most recent changes and consequently the changes would not be evaluated. Another limitation to be considered concerns the assumptions made by the team or analyst. The PHA is dependent on good judgment and the assumptions made during the study need to be documented and understood by the team and reviewer and kept for a future PHA.

3. PHA Team

The team conducting the PHA needs to understand the methodology that is going to be used. A PHA team can vary in size from two people to a number of people with varied operational and technical backgrounds. Some team members may only be a part of the team for a limited time. The team leader needs to be fully knowledgeable in the proper implementation of the PHA methodology that is to be used and should be impartial in the evaluation. The other full- or part-time team members need to provide the team with expertise in areas such as process technology, process design, operating procedures and practices, including how the work is actually performed, alarms, emergency procedures, instrumentation, maintenance procedures, both routine and nonroutine tasks, including how the tasks are authorized, procurement of parts and supplies, safety and health, and any other relevant subject as the need dictates. At least one team member must be familiar with the process.

The ideal team will have an intimate knowledge of the standards, codes, specifications, and regulations applicable to the process being studied. The selected team members need to be compatible and the team leader needs to be able to manage the team and the PHA study. The team needs to be able to work together while benefiting from the expertise of others on the team or outside the team, to resolve issues, and to forge a consensus on the findings of the study and the recommendations.

4. PHA Application

The application of a PHA to a process may involve the use of different methodologies for various parts of the process. For example, a process involving a series of unit operations of varying sizes, complexities, and ages may use different methodologies and team members for each operation. Then the conclusions can be integrated into one final study and evaluation. A more specific example is the use of a checklist PHA for a standard boiler or heat exchanger and the use of a Hazard and Operability PHA for the overall process. Also, for batch type processes like custom batch operations, a generic PHA of a representative batch may be used where there are only small changes of monomer or other ingredient ratios and the chemistry is documented for the full range and ratio of batch ingredients. Another process that might consider using a generic type of PHA is a gas plant. Often these plants are simply moved from site to site and therefore, a generic PHA may be used for these movable plants. Also, when an employer has several similar size gas plants and no sour gas is being processed at the site, then a generic PHA is feasible as long as the variations of the individual sites are accounted for in the PHA. Finally, when an employer has a large continuous process which has several control rooms for different portions of the process such as for a distillation tower and a blending operation, the employer may wish to do each segment separately and then integrate the final results.

Additionally, small businesses which are covered by this rule, will often have processes that have less storage volume, less capacity, and are less complicated than processes at a large facility. Therefore, OSHA would anticipate that the less complex methodologies would be used to meet the process hazard analysis criteria in the standard. These process hazard analyses can be done in less time and with only a few people being involved. A less complex process generally means that less data, P&IDs, and process information are needed to perform a process hazard analysis.

Many small businesses have processes that are not unique, such as cold storage lockers or water treatment facilities. Where employer associations have a number of members with such facilities, a generic PHA, evolved from a checklist or what-if questions, could be developed and used by each employer effectively to reflect his/her particular process; this would simplify compliance for them.

5. PHA Priority

When the employer has a number of processes which require a PHA, the employer must set up a priority system of which PHAs to conduct first. A preliminary or gross hazard analysis may be useful in prioritizing the processes that the employer has determined are subject to coverage by the process safety management standard. Consideration should first be given to those processes with the potential of adversely affecting the largest number of employees. This prioritizing should consider the potential severity of a chemical release, the number of potentially affected employees, the operating history of the process such as the frequency of chemical releases, the age of the process, and any other relevant factors. These factors would suggest a ranking order and would suggest either using a weighing factor system or a systematic ranking method. The use of a preliminary hazard analysis would assist an employer in determining which process should be of the highest priority and thereby the employer would obtain the greatest improvement in safety at the facility.

6. Additional Guidance

OSHA suggests that engineering controls include: appropriate application of detection methodologies to provide early warning of release, substitution of less hazardous materials, protective systems such as deluges, monitors, foams, increased separation distances, modification of the process temperature or pressure, and redundancy in instrumentation.

In regard to facility siting, OSHA looks for calculations, charts and other documents that verify facility siting has been considered. For example, safe distances for locating control rooms may be based on studies of the individual characteristics of equipment involved such as: types of construction of the room, types and quantities of materials, types of reaction and processes, operating pressures and temperature, presence of ignition sources, fire protection facilities, capabilities to respond to explosions, drainage facilities, and location of fresh air intakes.

Human factors may include a review of operator/process and operator/equipment interface, the number of tasks operators must perform and the frequency, the evaluation of extended or unusual work schedules, the clarity and simplicity of control displays, automatic instrumentation versus manual procedures, operator feedback, and clarity of signs and codes.

OSHA expects to see hydrocarbon or toxic gas monitors and alarms, electrical classifications consistent with flammability hazards, destruct systems such as flares in place and operating, control room siting adequate or provisions made for blast resistant construction, pressurization, or alarms, all relief valves and rupture discs properly designed and discharging to a safe area, and pipework protected from impact.

C. TYPES OF HAZARDS ANALYSIS TECHNIQUES[4]
1. What-if/Checklist

What-if analysis involves brainstorming by a group of process safety professionals to develop questions related to the hazards, hazardous situations, or specific accidents related to a process. The questions are written down, grouped into related areas such as maintenance or operations, and then given to experts in those areas to address. Once the questions are answered, the team gets together to review and discuss the results. In considering a potential accident, the team considers the incident scenario, identifies the potential consequences, qualitatively assesses the seriousness of the consequences, evaluates their likelihood, and make recommendations to mitigate the hazards.

A problem with the what-if approach is that is relatively unrestrained and can develop questions that are too far-reaching and broad. On the other hand, a simple checklist approach is too constrained, and potential hazards and concerns can be overlooked. The what-if/checklist approach combines the freewheeling what-if analysis with the structured checklist to combine the strengths of both approaches. This technique is commonly used in the initial stages of a comprehensive hazards analysis.

2. Hazard and Operability Study (HAZOP)

This technique is widely used in the chemical industry to evaluate process safety, environmental hazards, and processing problems. Its intent is to systematically uncover all credible process deviations in a process and evaluate the consequences of each. In a HAZOP analysis, a process knowledgeable team examines each segment of a process using a standard set of guide words to identify all significant deviations from normal conditions. Once the deviations are identified, the team identifies how they might occur, and then accounts for their detection and mitigation. When unresolved problems are identified, recommendations are made for corrective action.

The HAZOP analysis process has proven useful for a wide range of problems. It can be used for continuous and batch operations, it can be used to evaluate operating procedures and potential human errors, and it can consider multiple equipment errors.

3. Failure Mode and Effects Analysis (FMEA)

In this technique the various failure modes of equipment in a process are delineated (using the guidance in a checklist) and then the potential effects on the system or plant are evaluated. Equipment failures can initiate or contribute to accidents or they might not. While this analysis technique is methodical and straightforward, it has deficiencies in that it does not consider combinations of equipment failures and does not include human errors.

4. Fault-Tree Analysis (FTA)

Fault-tree analysis is a means of analyzing rather than identifying accidents. It is a graphical deductive technique that focuses on one type of accident or system failure and displays the various primary and secondary causes (equipment failures, human errors, external factors). The strengths of this method are its abilities to identify combinations of events that can lead to accidents and to evaluate the relative importance of the basic causes. Preventive actions can then be focused on the most significant problem causes.

5. Other Accident Analysis Methods

Among other accident analysis methods are Event-Tree Analysis, Cause-Consequence Analysis, and Human Reliability Analysis. In event-tree analysis, a specific initiating event is followed by other failures or events, until the final consequences are reached. The event-tree shows all the possible outcomes of the single initiating event, following through all the potential mitigation and protective measures and their potential outcomes. In cause-consequence analysis, the fault-tree and event-tree analysis techniques are combined and the relationships between accident consequences and their causes are displayed. This technique is most commonly used when the accident's failure logic is relatively simple. Human reliability analysis aims to systematically evaluate the factors that influence how operators, technicians, and other staff perform. It endeavors to identify situations that can lead to accidents and trace the effects of human error.

PROCESS HAZARDS ANALYSIS: ASSESSMENT QUESTIONS

- Has there been a process hazards analysis (PHA) appropriate to the complexity and hazards of the process?
- Does the PHA address:
  - process hazards?
  - previous incidents?
  - failures of administrative and engineering controls?
  - facility siting?
  - human factors?
- Was a standard PHA methodology employed?
- Was the PHA methodology employed by a team knowledgeable of:
  - engineering?
  - process operations?
  - the PHA methodology used?
- Were any negative PHA results promptly corrected?
- Were the PHA results communicated to all affected employees?
- Is the PHA updated every five years?
- Are all PHA's and records of recommendation resolution kept for the life of the process?

V. OPERATING PROCEDURES AND PRACTICES

A. OSHA REQUIREMENTS

The OSHA standard requires that:

(1) The employer shall develop and implement written operating procedures that provide clear instructions for safely conducting activities involved in each . . . process consistent with the process safety information and shall address at least the following elements.

    (i) Steps for each operating phase:
        (A) Initial startup;
        (B) Normal operations;
        (C) Temporary operations;
        (D) Emergency shutdown including the conditions under which emergency shutdown is required, and the assignment of shutdown responsibility to qualified operators to ensure that emergency shutdown is executed in a safe and timely manner.
        (E) Emergency Operations;
        (F) Normal shutdown; and,
        (G) Startup following a turnaround, or after an emergency shutdown.
    (ii) Operating limits:
        (A) Consequences of deviation; and
        (B) Steps required to correct or avoid deviation.
    (iii) Safety and health considerations:
        (A) Properties of, and hazards presented by, the chemicals used in the process;
        (B) Precautions necessary to prevent exposure, including engineering controls, administrative controls, and personal protective equipment;
        (C) Control measures to be taken if physical contact or airborne exposure occurs; *Control primarily means first aid procedures or emergency medical attention, which should be consistent with the information on the MSDS.*
        (D) Quality control for raw materials and control of hazardous chemical inventory levels; and,
        (E) Any special or unique hazards.
    (iv) Safety systems and their functions.
OSHA believes that tasks and procedures related to the covered process must be appropriate, clear, consistent, and most importantly, well communicated to employees.

(2) Operating procedures shall be readily accessible to employees who work in or maintain a process. *To ensure that a ready and up-to-date reference is available, and to form a foundation for need employee training.*

(3) The operating procedures shall be reviewed as often as necessary to assure that they reflect current operating practice, including changes that result from changes in process chemicals, technology, and equipment, and changes to facilities. The employer shall certify annually that these operating procedures are current and accurate.

(4) The employer shall develop and implement safe work practices to provide for the control of hazards during operations such as lockout/tagout; confined space entry; opening process equipment or piping; and control over entrance into a facility by maintenance, contractor, laboratory, or other support personnel. These safe work practices shall apply to employees and contractor employees.

B. BACKGROUND
1. Procedure Uses

OSHA's background on operating procedures and practices states that: Operating procedures describe tasks to be performed, data to be recorded, operating conditions to be maintained, samples to be collected, and safety and health precautions to be taken. The procedures need to be technically accurate, understandable to employees, and revised periodically to ensure that they reflect current operations. The process safety information package is to be used as a resource to better assure that the operating procedures and practices are consistent with the known hazards of the chemicals in the process and that the operating parameters are accurate. Operating procedures should be reviewed by engineering staff and operating personnel to ensure that they are accurate and provide practical instructions on how to actually carry out job duties safely.

2. Procedure Content

Operating procedures will include specific instructions or details on what steps are to be taken or followed in carrying out the stated procedures. These operating instructions for each procedure should include the applicable safety precautions and should contain appropriate information on safety implications. For example, the operating procedures addressing operating parameters will contain operating instructions about pressure limits, temperature ranges, flow rates, what to do when an upset condition occurs, what alarms and instruments are pertinent if an upset condition occurs, and other subjects. Another example of using operating instructions to properly implement operating procedures is in starting up or shutting down the process. In these cases, different parameters will be required from those of normal operation. These operating instructions need to clearly indicate the distinctions between startup and normal operations such as the appropriate allowances for heating up a unit to reach the normal operating parameters. Also the operating instructions need to describe the proper method for increasing the temperature of the unit until the normal operating temperature parameters are achieved.

Computerized process control systems add complexity to operating instructions. These operating instructions need to describe the logic of the software as well as the relationship between the equipment and the control system; otherwise, it may not be apparent to the operator.

3. Training Use of Procedures

Operating procedures and instructions are important for training operating personnel. The operating procedures are often viewed as the standard operating practices (SOPs) for operations. Control room personnel and operating staff, in general, need to have a full understanding of operating procedures. If workers are not fluent in English then procedures and instructions need to be prepared in a second language understood by the workers. In addition, operating procedures need to be changed when there is a change in the process as a result of the management of change procedures. The consequences of operating procedure changes need to be fully evaluated and the information conveyed to the personnel. For example, mechanical changes to the process made by the maintenance department (like changing a valve from steel to brass or other subtle changes) need to be evaluated to determine if operating procedures and practices also need to be changed. All management of change actions must be coordinated and integrated with current operating procedures and operating personnel must be oriented to the changes in procedures before the change is made. When the process is shutdown in order to make a change, then the operating procedures must be updated before startup of the process.

Training in how to handle upset conditions must be accomplished as well as what operating personnel are to do in emergencies such as when a pump seal fails or a pipeline ruptures. Communication between operating personnel and workers performing work within the process area, such as nonroutine tasks, also must be maintained. The hazards of the tasks are to be conveyed to operating personnel in accordance with established procedures and to those performing the actual tasks. When the work is completed, operating personnel should be informed to provide closure on the job.

OPERATING PROCEDURES AND PRACTICES: ASSESSMENT QUESTIONS

- Are there clear written process operating procedures for all phases of operations which address:
  - safety systems?
  - precautions?
  - limits?
  - safety and health considerations?
- Are the procedures reviewed and certified correct at least annually?
- Are there safe work practices for:
  - lockout/tagout?
  - confined space entry?
  - opening process equipment?
  - facility entry?
- Are operating procedures readily accessible?
- Are operating procedures used for training?
- Does procedure training include upset conditions and emergencies?

VI. EMPLOYEE TRAINING

A. OSHA REQUIREMENTS

The OSHA standard requires that:

(1) Initial training.

(i) Each employee presently involved in operating a process, and each employee before being involved in operating a newly assigned process, shall be trained in an overview of the process and in the operating procedures . . . . The training shall include emphasis on the specific safety and health hazards, emergency operations including shutdown, and safe work practices applicable to the employee's job tasks.

(ii) In lieu of initial training for those employees already involved in operating a process on . . . , an employer may certify in writing that the employee has the required knowledge, skills, and abilities to safely carry out the duties and responsibilities as specified in the operating procedures.

When new procedures are developed, the employer must give training to operating employees prior to their implementation.

(2) Refresher training. Refresher training shall be provided at least every three years, and more often if necessary, to each employee involved in operating a process to assure that the employee understands and adheres to the current operating procedures of the process. The employer, in consultation with the employees involved in operating the process, shall determine the appropriate frequency of refresher training.

The time period extends from the date of the last training or the "grandfathering." More frequent training should be based on factors such as deviations from standard operating procedures, recent incidents, or apparent deficiencies in training.

(3) Training documentation. The employer shall ascertain that each employee involved in operating a process has received and understood the training required . . . . The employer shall prepare a record which contains the identity of the employee, the date of training, and the means used to verify that the employee understood the training.

OSHA requires some positive means be taken by the employer to determine if employees have understood their training and are capable of adhering to current operating procedures. This could include the administration of a written test. Other means of ascertaining comprehension of the training, such as on-the-job demonstration, are acceptable as long as they are adequately documented.

B. BACKGROUND

1. Scope of Training

OSHA's background on employee training states that: All employees, including maintenance and contractor employees, involved with highly hazardous chemicals need to fully understand the safety and health hazards of the chemicals and processes they work with for the protection of themselves, their fellow employees, and the citizens of nearby communities. Training conducted in compliance with 1910.1200, the Hazard Communication standard, will help employees to be more knowledgeable about the chemicals they work with as well as familiarize them with reading and understanding MSDS. However, additional training in subjects such as operating procedures and safety work practices, emergency evacuation and response, safety procedures, routine and nonroutine work authorization activities, and other areas pertinent to process safety and health will need to be covered by an employer's training program.

In establishing their training programs, employers must clearly define the employees to be trained and what subjects are to be covered in their training. Employers, in setting up their training program, will need to clearly establish the goals and objectives they wish to achieve with the training that they provide to their employees. The learning goals or objectives should be written in clear measurable terms before the training begins. These goals and objectives need to be tailored to each of the specific training modules or segments. Employers should describe the important actions and conditions under which the employee will demonstrate competence or knowledge as well as what is acceptable performance.

2. Hands-On Training

Hands-on-training where employees are able to use their senses beyond listening, will enhance learning. For example, operating personnel, who will work in a control room or at control panels, would benefit by being trained at a simulated control panel or panels. Upset conditions of various types could be displayed on the simulator, and then the employee could go through the proper operating procedures to bring the simulator panel back to the normal operating parameters. A training environment could be created to help the trainee feel the full reality of the situation but, of course, under controlled conditions. This realistic type of training can be very effective in teaching employees correct procedures while allowing them to also see the consequences of what might happen if they do not follow established operating procedures. Other training techniques using videos or on-the-job training can also be very effective for teaching other job tasks, duties, or other important information. An effective training program will allow the employee to fully participate in the training process and to practice their skill or knowledge.

3. Training Evaluation

Employers need to periodically evaluate their training programs to see if the necessary skills, knowledge, and routines are being properly understood and implemented by their trained employees. The means or methods for evaluating the training should be developed along with the training program goals and objectives. Training program evaluation will help employers to determine the amount of training their employees understood, and whether the desired results were obtained. If, after the evaluation, it appears that the trained employees are not at the level of knowledge and skill that was expected, the employer will need to revise the training program, provide retraining, or provide more frequent refresher training sessions until the deficiency is resolved. Those who conducted the training and those who received the training should also be consulted as to how best to improve the training process. If there is a language barrier, the language known to the trainees should be used to reinforce the training messages and information.

4. Retraining

Careful consideration must be given to assure that employees including maintenance and contract employees receive current and updated training. For example, if changes are made to a process, impacted employees must be trained in the changes and understand the effects of the changes on their job tasks (e.g., any new operating procedures pertinent to their tasks). Additionally, as already discussed the evaluation of the employee's absorption of training will certainly influence the need for training.

The Center for Chemical Process Safety notes that retraining is often required because of changes in staff and promotions. Also there is a common "drifting away" from strict adherence to procedures.[5] In such cases the following techniques have been found useful to prevent memory loss.
- using incident investigations in training;
- periodic publicity reminders of past accidents;
- reviewing procedures to evaluate their relevance and accuracy;
- walking the plant to see if short cuts are being taken;
- asking personnel why they are doing the jobs in certain ways; and
- seeing if work permits are accurate and are being implemented as required.

EMPLOYEE TRAINING: ASSESSMENT QUESTIONS

- Are all process employees appropriately trained in process safety?
- Does training include:
  - all operating phase steps?
  - operating limits?
  - safety and health considerations?
  - safety systems and their functions?
- Is training planned with clear goals and objectives?
- Are process operators retrained at least every three years?
- Does each employee have a training file?
- Is training competence assured?
- Do training techniques include hands-on, simulators, videos, etc.?
- Is there a comprehensive training program review?
- For experienced employees, has a certification been used in lieu of training?

VII. CONTRACTORS

A. OSHA REQUIREMENTS

The OSHA standard requires that:

(1) Application. This paragraph applies to contractors performing maintenance or repair, turnaround, major renovation, or specialty work on or adjacent to a covered process. It does not apply to contractors providing incidental services which do not influence process safety . . . .

(2) Employer responsibilities.

(i) The employer . . . selecting a contractor, shall obtain and evaluate information regarding the contract employer's safety performance and programs.

This also applies to subcontractors. The employer and general contractor are both responsible for ensuring that the duties in this paragraph are performed; this applies to inquiring into the safety records of their subcontractors, informing the subcontractor of the known potential hazards, the emergency action plan, and safe work practices, and ensuring the subcontractor's compliance with the standard. Also the employer has the obligation to ensure that the contract employer and the subcontractor are properly performing their obligations with respect to the subcontractor's compliance with the standard. Host employers and contractors should exercise responsible oversight of their respective contractors' and subcontractors' performance of safety and health requirements of the standard.

(ii) The employer shall inform contract employers of the known potential fire, explosion, or toxic release hazards related to the contractor's work and the process.

(iii) The employer shall explain to contract employers the applicable provisions of the emergency action plan . . . .

(iv) The employer shall develop and implement safe work practices . . . to control the entrance, presence, and exit of contract employers and contract employees in covered process areas.

(v) The employer shall periodically evaluate the performance of contract employers in fulfilling their obligations . . . .

(vi) The employer shall maintain a contract employee injury and illness log related to the contractor's work in process areas.

In terms of documentation, OSHA accepts the employer sharing contractor's OSHA 200 and 101 reports or equivalent, if those logs and reports specifically indicate which injuries and illnesses are related to process areas.

(3) Contract employer responsibilities.

(i) The contract employer shall assure that each contract employee is trained in the work practices necessary to safely perform his/her job.

(ii) The contract employer shall assure that each contract employee is instructed in the known potential fire, explosion, or toxic release hazards related to his/her job and the process, and the applicable provisions of the emergency action plan.

(iii) The contract employer shall document that each contract employee has received and understood the training required . . . . The contract employer shall prepare a record which contains the identity of the contract employee, the date of training, and the means used to verify that the employee understood the training.

(iv) The contract employer shall assure that each contract employee follows the safety rules of the facility including the safe work practices required . . . .

(v) The contract employer shall advise the employer of any unique hazards presented by the contract employer's work, or of any hazards found by the contract employer's work.

B. BACKGROUND
1. Contractor Screening
OSHA's background on contractors states that: Employers who use contractors to perform work in and around processes that involve highly hazardous chemicals, will need to establish a screening process so that they hire and use contractors who accomplish the desired job tasks without compromising the safety and health of employees at a facility. For contractors, whose safety performance on the job is not known to the hiring employer, the employer will need to obtain information on injury and illness rates and experience and should obtain contractor references.

Additionally, the employer must assure that the contractor has the appropriate job skills, knowledge and certifications (such as for pressure vessel welders). Contractor work methods and experiences should be evaluated. For example, does the contractor conducting demolition work swing loads over operating processes or does the contractor avoid such hazards?

2. Injury and Illness Logs

Maintaining a site injury and illness log for contractors is another method employers must use to track and maintain current knowledge of work activities involving contract employees working on or adjacent to covered processes. Injury and illness logs of both the employer's employees and contract employees allow an employer to have full knowledge of process injury and illness experience. This log will also contain information which will be of use to those auditing process safety management compliance and those involved in incident investigations.

3. Work Control

Contract employees must perform their work safely. Since contractors often perform very specialized and potentially hazardous tasks such as confined space entry activities and nonroutine repair, their activities need to be controlled while they are working on or near a covered process. A permit system or work authorization system for these activities would also be helpful to all affected employers. A work authorization system keeps an employer informed of contract employee activities, and gives the employer more management control over the work being performed. A well-run and well-maintained process where employee safety is fully recognized will benefit all of those who work in the facility whether they be contract employees or employees of the owner.

4. Contractor Training

The burden of training contractor employees is on the contract employer. But, the host employer bears the responsibility to periodically evaluate the performance of contract employers in fulfilling their obligations as specified. This clearly includes training obligations.

If contract employees are involved in operating a process or maintaining . . . process equipment, then they must receive training as specified . . . . The host employer must ensure, through periodic evaluations, that the training provided is equivalent to the training required for direct hire employees. Such training need not be identical in format or content or context to training given to the host's employees. The critical element is that the information required by the standard must be conveyed to and learned by the contract employees . . . .

Contract employees must be able to perform their own job tasks safely and should receive: training prior to beginning work on or near a covered process which should encompass instruction regarding known process hazards related to their jobs including training in the applicable provisions of the emergency action plan, and training in the safe work practices adopted by the host employer and the contract employer. They should also receive additional training as necessary to prepare the workers for changes in operations or work practices at the facility and to ensure that the employee's understanding of the applicable safe work practices and other rules remains current.

CONTRACTORS: ASSESSMENT QUESTIONS

- Are contract employers' safety records reviewed prior to hire?
- Are all hazards and safety provisions (including emergency plans) described to the contract employers and employees?
- Are provisions in place to control and monitor contract work?
- Are all contract employees appropriately trained (equivalent to direct hires)?
- Is all contract training ensured and documented?
- Is a contract employee injury and illness log for the contractor's work in process areas maintained?
- Is contract work periodically reviewed?

VIII. PRE-STARTUP SAFETY

A. OSHA REQUIREMENTS
The OSHA standard requires that:

(1) The employer shall perform a pre-startup safety review for new facilities and for modified facilities when the modification is significant enough to require a change in the process safety information.

(2) The pre-startup safety review shall confirm that prior to the introduction of highly hazardous chemicals to a process:

(i) Construction and equipment is in accordance with design specifications;

For equipment that has been modified to the extent that a change to the process safety information is required, the employer must ensure that the process safety information has been modified prior to startup.

(ii) Safety, operating, maintenance, and emergency procedures are in place and are adequate;

(iii) For new facilities, a process hazard analysis has been performed and recommendations have been resolved or implemented before startup; and modified facilities meet the requirements contained in management of change . . . .

(iv) Training of each employee involved in operating a process has been completed.

B. BACKGROUND
1. New Processes
OSHA's background on pre-startup safety states that: For new processes, the employer will find a PHA helpful in improving the design and construction of the process from a reliability and quality point of view. The safe operation of the new process will be enhanced by making use of the PHA recommendations before final installations are completed. P&IDs are to be completed along with having the operating procedures in place and the operating staff trained to run the process before startup. The initial startup procedures and normal operating procedures need to be fully evaluated as part of the pre-startup review to assure a safe transfer into the normal operating mode for meeting the process parameters.

2. Existing Processes
For existing processes that have been shutdown for turnaround, or modification, etc., the employer must assure that any changes other than "replacement in kind" made to the process during shutdown go through the management of change procedures. P&IDs will need to be updated as

necessary, as well as operating procedures and instructions. If the changes made to the process during shutdown are significant and impact the training program, then operating personnel as well as employees engaged in routine and nonroutine work in the process area may need some refresher or additional training in light of the changes. Any incident investigation recommendations, compliance audits, or PHA recommendations need to be reviewed as well to see what impacts they may have on the process before beginning the startup.

PRE-STARTUP SAFETY: ASSESSMENT QUESTIONS

- Is there a comprehensive pre-startup safety review for new or modified facilities that includes:
  - design and construction?
  - procedures?
  - a hazards analysis?
- For new processes are all procedures in place and employees trained prior to startup?
- Have process startup procedures been carefully reviewed?
- Do changes to existing procedures go through the management of change system?
- Are employees appropriately trained in changed procedures?

IX. MECHANICAL INTEGRITY

A. OSHA REQUIREMENTS
The OSHA standard requires that:

(1) Application. *(The mechanical integrity requirements)* apply to the following process equipment:

  (i)   Pressure vessels and storage tanks;
  (ii)   Piping systems (including piping components such as valves);
  (iii)   Relief and vent systems and devices;
  (iv)   Emergency shutdown systems;
  (v)   Controls (including monitoring devices and sensors, alarms, and interlocks) and,
  (vi)   Pumps.

(2) Written procedures. The employer shall establish and implement written procedures to maintain the on-going integrity of process equipment.

OSHA believes that the written procedures must be in adequate detail to ensure that the specific process equipment receives careful, appropriate, regularly scheduled maintenance to ensure its continued safe operation. In specific, a "breakdown" maintenance program does not meet the requirements specified.

Also, the written procedures need to be specific to the type of vessel or equipment. Identical or very similar vessels and items of equipment in similar service need not have individualized maintenance procedures. Each procedure must clearly identify the equipment to which it applies.

(3) Training for process maintenance activities. The employer shall train each employee involved in maintaining the on-going integrity of process equipment in an overview of that process and its hazards and in the procedures applicable to the employee's job tasks to assure that the employee can perform the job tasks in a safe manner.

OSHA wants ongoing or continual training to ensure that employees can perform their jobs in a safe manner. New maintenance employees should be trained before beginning work at the site and all maintenance employees should receive additional training appropriate to their constantly changing job tasks.

Although maintenance employees should not be trained in process operating procedures to the same extent as those employees who are actually involved in operating the process, they must be trained in all "procedures applicable to the employee's job tasks to assure that the employee can perform the job tasks in a safe manner." Thus, a maintenance worker sent to work on a process breakdown must be trained in operating procedures that are relevant to the repair or installation on which he or she is working.

Employers should incorporate all safety-related topics applicable to maintenance tasks into the ongoing training program required to assure that maintenance employees can perform their tasks in a safe manner. Thus, under appropriate circumstances, the employer must train these workers in the safe work practices required, the written procedures to manage change, and in appropriate provisions of the emergency action plan.

(4) Inspection and testing.

(i) Inspections and tests shall be performed on process equipment.

(ii) Inspection and testing procedures shall follow recognized and generally accepted good engineering practices.

(iii) The frequency of inspections and tests of process equipment shall be consistent with applicable manufacturers' recommendations and good engineering practices, and more frequently if determined to be necessary by prior operating experience.

(iv) The employer shall document each inspection and test that has been performed on process equipment. The documentation shall identify the date of the inspection or test, the name of the person who performed the inspection or test, the serial number or other identifier of the equipment on which the inspection or test was performed, a description of the inspection or test performed, and the results of the inspection or test.

(5) Equipment deficiencies. The employer shall correct deficiencies in equipment that are outside acceptable limits . . . before further use or in a safe and timely manner when necessary means are taken to assure safe operation.

OSHA expects equipment deficiencies to be corrected promptly if the equipment is outside the acceptable limits specified in the process safety information. There may be situations where it may not be necessary that the deficiencies be corrected "before further use" as long as the deficiencies are corrected in a timely and safe manner when necessary means (e.g., protective measures and continuous monitoring) are taken to ensure safe operation.

(6) Quality assurance.

(i) In the construction of new plants and equipment, the employer shall assure that equipment as it is fabricated is suitable for the process application for which they will be used.

(ii) Appropriate checks and inspections shall be performed to assure that equipment is installed properly and is consistent with design specifications and the manufacturer's instructions.

(iii) The employer shall assure that maintenance materials, spare parts, and equipment are suitable for the process application for which they will be used.

B. BACKGROUND

1. The Mechanical Integrity Program

OSHA's background on mechanical integrity states that: Employers will need to review their maintenance programs and schedules to see if there are areas where "breakdown" maintenance is used rather than an on-going mechanical integrity program. Equipment used to process, store, or handle highly hazardous chemicals needs to be designed, constructed, installed, and maintained to minimize the risk of releases of such chemicals. This requires that a mechanical integrity program be in place to assure the continued integrity of process equipment. Elements of a mechanical integrity program include the identification and categorization of equipment and instrumentation, inspections and tests, testing and inspection frequencies, development of maintenance procedures, training of maintenance personnel, the establishment of criteria for acceptable test results, documentation of test and inspection results, and documentation of manufacturer recommendations as to mean time to failure for equipment and instrumentation.

2. Lines of Defense

The first line of defense an employer has available is to operate and maintain the process as designed, and to keep the chemicals contained. This line of defense is backed up by the next line of defense which is the controlled release of chemicals through venting to scrubbers or flares, or to surge or overflow tanks which are designed to receive such chemicals, etc. These lines of defense are the primary lines of defense or means to prevent unwanted releases. The secondary lines of defense would include fixed fire protection systems like sprinklers, water spray, or deluge systems, monitor guns, etc., dikes, designed drainage systems, and other systems which would control or mitigate hazardous chemicals once an unwanted release occurs. These primary and secondary lines of defense are what the mechanical integrity program needs to protect and strengthen these primary and secondary lines of defenses where appropriate.

3. Equipment List and Schedule

The first step of an effective mechanical integrity program is to compile and categorize a list of process equipment and instrumentation for inclusion in the program. This list would include pressure vessels, storage tanks, process piping, relief and vent systems, fire protection system components, emergency shutdown systems and alarms and interlocks and pumps. For the categorization of instrumentation and the listed equipment, the employer would prioritize which pieces of equipment require closer scrutiny than others. Mean time to failure of various instrumentation and equipment parts would be known from the manufacturers' data or the employer's experience with the parts, which would then influence the inspection and testing frequency and associated procedures. Also, applicable codes and standards such as the National Board Inspection Code, or those from the American Society for Testing and Material, American Petroleum Institute, National Fire Protection Association, American National Standards Institute, American Society of Mechanical Engineers, and other groups, provide information to help establish an effective testing and inspection frequency, as well as appropriate methodologies.

4. Inspection Criteria

The applicable codes and standards provide criteria for external inspections for such items as foundation and supports, anchor bolts, concrete or steel supports, guy wires, nozzles and sprinklers, pipe hangers, grounding connections, protective coatings and insulation, and external metal surfaces of piping and vessels, etc. These codes and standards also provide information on methodologies for internal inspection, and a frequency formula based on the corrosion rate of the materials of construction. Also, erosion both internal and external, needs to be considered along with corrosion effects for piping and valves. Where the corrosion rate is not known, a maximum inspection frequency is recommended, and methods of developing the corrosion rate are available in the codes. Internal inspections need to cover items such as vessel shell, bottom, and head; metallic linings; nonmetallic linings; thickness measurements for vessels and piping; inspection for erosion, corrosion, cracking, and bulges; internal equipment like trays, baffles, sensors, and screens for erosion, corrosion or cracking, and other deficiencies. Some of these inspections may be performed by state or local government inspectors under state and local statutes. However, each employer needs to develop procedures to ensure that tests and inspections are conducted properly and that consistency is maintained even where different employees may be involved. Appropriate training is to be provided to maintenance personnel to ensure that they understand the preventive maintenance program procedures, safe practices, and the proper use and application of special equipment or unique tools that may be required. This training is part of the overall training program called for in the standard.

5. Quality Assurance

A quality assurance system is needed to help ensure that the proper materials of construction are used, that fabrication and inspection procedures are proper, and that installation procedures recognize field installation concerns. The quality assurance program is an essential part of the mechanical integrity program and will help to maintain the primary and secondary lines of defense that have been designed into the process to prevent unwanted chemical releases or those which control or mitigate a release. "As built" drawings, together with certifications of coded vessels and other equipment, and materials of construction need to be verified and retained in the quality assurance documentation. Equipment installation jobs need to be properly inspected in the field for use of proper materials and procedures and to assure that qualified craftsmen are used to do the job. The use of appropriate gaskets, packing, bolts, valves, lubricants, and welding rods need to be verified in the field. Also, procedures for installation of safety devices need to be verified, such as the torque on the bolts on ruptured disc installations, uniform torque on flange bolts, proper installation of pump seals, etc. If the quality of parts is a problem, it may be appropriate to conduct audits of the equipment supplier's facilities . . . . Any changes in equipment that may become necessary will need to go through the management of change procedures.

MECHANICAL INTEGRITY: ASSESSMENT QUESTIONS

- Is there a written equipment integrity program for:
  - pressure vessels?
  - piping systems?
  - reliefs?
  - emergency equipment?
  - control equipment?
- Is process equipment designed, constructed, installed, and maintained to minimize the risk of accidents?
- Has all equipment been reviewed, and is all equipment in the program listed?
- Are maintenance employees trained appropriately?
- Does equipment inspection and testing follow recognized engineering practices and manufacturer's recommendations for frequencies and technologies?
- Are all inspection and testing properly documented with:
  - date?
  - tester?
  - equipment identification?
  - test performed?
  - results?
- Are deficiencies appropriately corrected?
- Are maintenance materials and spare parts suitable for the process application?

X. NONROUTINE WORK AUTHORIZATIONS

OSHA states that the intent of this section is that employers shall control, in a consistent manner, nonroutine work conducted in process areas. Specifically, this section deals with the permitting of hot work operations associated with welding and cutting in process areas. Hot work permits need to comply with the OSHA's fire prevention and protection requirements of 29 CFR 1910.252 (a).

A. OSHA REQUIREMENTS
The OSHA standard states that:
(1) The employer shall issue a hot work permit for hot work operations conducted on or near a covered process.
(2) The permit shall document that..... fire prevention and protection requirements .... have been implemented prior to beginning the hot work operations; it shall indicate the date(s) authorized for hot work; and identify the object on which hot work is to be performed. The permit shall be kept on file until completion of the hot work operations.

B. BACKGROUND
1. General Guidance
OSHA's background on nonroutine work authorizations (hot work permits) states that: Nonroutine work which is conducted in process areas needs to be controlled by the employer in a consistent manner. The hazards identified involving the work that is to be accomplished must be communicated to those doing the work, but also to those operating personnel whose work could affect the safety of the process. A work authorization notice or permit must have a procedure that describes the steps the maintenance supervisor, contractor representative, or other person needs to follow to obtain the necessary clearance to get the job started. The work authorization procedures need to reference and coordinate, as applicable, lockout/tagout procedures, line breaking procedures, confined space entry procedures, and hot work authorizations. This procedure also needs to provide clear steps to follow once the job is completed in order to provide closure for those that need to know the job is now completed and equipment can be returned to normal.

2. Hot Work Permit Requirements
OSHA's requirements mandate that:
- *Hot work permits identify openings, cracks, holes where sparks may drop to combustible materials below.*
- *Hot work permits describe the fire extinguishers required to handle any emergencies.*
- *Hot work permits assign fire watchers where other than a minor fire might develop.*
- *Hot work permits are authorized, preferably in writing, by the individual responsible for all welding and cutting operations. The authorization is preceded by site inspection and designation of appropriate precautions.*
- *Hot work permits describe precautions associated with combustible materials on floors or floors, walls, roofs, partitions, ceilings of combustible construction.*

3. Hot Work Permit Compliance
In looking at compliance, OSHA also seeks to find:
- *Has hot work permitting been successful in prohibiting welding in unauthorized areas, in sprinklered buildings while such protection is impaired, in the presence of explosive atmospheres, and in storage areas for large quantities of readily ignitable materials?*

- *Have the hot work permits required the relocation of combustibles where practicable and covering with flame proofed covers where not practicable?*
- *Have hot work permits identified for shutdown any ducts or conveyor systems that may convey sparks to distant combustibles?*
- *Have hot work permits required precautions, whenever welding on components that could transmit heat by radiation or conduction to unobserved combustibles?*
- *Have hot work permits identified hazards associated with welding on walls, partitions, ceilings, or roofs with combustible coverings or welding on walls or panels of sandwich type construction?*
- *Has management established areas and procedures for safe welding and cutting based on fire potential?*
- *Has management designated the individual responsible for authorizing cutting and welding operations in process areas?*
- *Has management ensured that welders, cutters, and supervisors are trained in the safe operations of their equipment?*
- *Has management advised outside contractors working on their site about all hot work permitting programs?*
- *Has the supervisor determined if combustibles are being protected from ignition prior to welding by moving them, shielding them, or scheduling welding around their production?*
- *Has the supervisor, prior to welding, secured authorization from his responsible individual designated by management?*

NONROUTINE WORK AUTHORIZATIONS: ASSESSMENT QUESTIONS

- Is there a hot work permit for hot work operations conducted on or near a potentially dangerous process?
- Does the permit document the fire prevention and protection requirements? the date(s) authorized for hot work? and the object on which hot work is to be performed?
- Has management ensured the proper training of the individuals and supervisors involved in the hot work?
- Is the permit kept on file until completion of the hot work?
- Is there a work authorization procedure for lockout/tagout? line breaking? and confined space entry?

XI. MANAGING CHANGE

The intent of these provisions is to require management of all modifications to equipment, procedures, raw materials, and processing conditions other than replacement in kind by identifying them and reviewing them prior to the implementation of the change. Management of change provisions are triggered by any change whatsoever that may affect a covered process except replacement in kind.

A. OSHA REQUIREMENTS
The OSHA standard requires that:
(1) The employer shall establish and implement written procedures to manage changes (except for "replacements in kind") to process chemicals, technology, equipment, and procedures; and, changes to facilities that affect a covered process.

(2) The procedures shall assure that the following considerations are addressed prior to any change:

    (i)   The technical basis for the proposed change;

    (ii)   Impact of change on safety and health;

    (iii)   Modifications to operating procedures;

    (iv)   Necessary time period for the change; and,

    (v)   Authorization requirements for the proposed change.

(3) Employees involved in operating a process and maintenance and contract employees whose job tasks will be affected by a change in the process shall be informed of, and trained in, the change prior to startup of the process or affected part of the process.

(4) If a change covered by this paragraph results in a change in the process safety information . . . such information shall be updated accordingly.

(5) If a change covered by this paragraph results in a change in the operating procedures or practices . . . such procedures or practices shall be updated accordingly.

B. BACKGROUND

1. Purpose of Change Management

OSHA's background on managing change states that: To properly manage changes to process chemicals, technology, equipment, and facilities, one must define what is meant by change. In this process safety management standard, change includes all modifications to equipment, procedures, raw materials, and processing conditions other than "replacement in kind." These changes need to be properly managed by identifying and reviewing them prior to implementation of the change. For example, the operating procedures contain the operating parameters (pressure limits, temperature ranges, flow rates, etc.) and the importance of operating within these limits. While the operator must have the flexibility to maintain safe operation within the established parameters, any operation outside of these parameters requires review and approval by a written management of change procedure.

2. Scope

Management of change covers changes in process technology and changes to equipment and instrumentation. Changes in process technology can result from changes in production rates, raw materials, experimentation, equipment unavailability, new equipment, new product development, change in catalyst, and changes in operating conditions to improve yield or quality. Equipment changes include, among others, change in materials of construction, equipment specifications, piping pre-arrangements, experimental equipment, computer program revisions, and changes in alarms and interlocks. Employers need to establish means and methods to detect both technical changes and mechanical changes.

3. Temporary Changes

Temporary changes have caused a number of catastrophes over the years, and employers need to establish ways to detect temporary changes as well as those that are permanent. It is important that a time limit for temporary changes be established and monitored since, without control, these changes may tend to become permanent. Temporary changes are subject to the management of change provisions. In addition, the management of change procedures are used to ensure that the equipment and procedures are returned to their original or designed conditions at the end of the temporary change. Proper documentation and review of these changes is invaluable in assuring that the safety and health considerations are being incorporated into the operating procedures and the process.

4. The Change Management Process

Employers may wish to develop a form or clearance sheet to facilitate the processing of changes through the management of change procedures. A typical change form may include a description and the purpose of the change, the technical basis for the change, safety and health considerations, documentation of changes for the operating procedures, maintenance procedures, inspection and testing, P&IDs, electrical classification, training and communications, pre-startup inspection, duration if a temporary change, approvals and authorization. Where the impact of the change is minor and well understood, a checklist reviewed by an authorized person with proper communication to others who are affected may be sufficient. However, for a more complex or significant design change, a hazard evaluation procedure with approvals by operations, maintenance, and safety departments may be appropriate. Changes in documents such as P&IDs, raw materials, operating procedures, mechanical integrity programs, electrical classifications, etc. need to be noted so that these revisions can be made permanent when the drawings and procedure manuals are updated. Copies of process changes need to be kept in an accessible location to ensure that design changes are available to operating personnel as well as to PHA team members when a PHA is being done or one is being updated.

C. PLANT IMPLEMENTATION

The Center for Chemical Process Safety lists the kinds of changes that can occur in chemical processes.[6] They include:

- changes in process chemicals and raw materials;
- changes in process technology such as changes in production rates, experimentation, and new product development;
- changes in equipment including materials of construction, specifications, and process systems;
- changes in instrumentation including computer programs, and alarms and interlocks;
- changes in operating procedures and practices;
- changes in facilities including buildings and containers that affect the process;
- changes in process conditions and operating parameters;
- and changes in personnel.

Formally, process change is defined as "a temporary or permanent substitution, alteration, replacement (excluding in kind replacements), or modification by adding or deleting process equipment, applicable codes, process control, catalysts or chemicals, feedstocks, mechanical procedures, electrical procedures, and safety procedures from the present configuration or the process equipment, procedures or operating limits."

In practice, implementing a management of change process at the operational level usually involves the following steps:

- a local definition of what process changes are;
- the process and mechanical design basis for the proposed change;
- the effects on interrelated upstream and downstream facilities;
- needed revisions to the operating procedures;
- necessary training of appropriate personnel;
- duration of the change;
- and a required authorization.

D. A CHEMICAL PLANT EXAMPLE

Jakubowski[7] describes the development and installation of a management of change program at a bulk pharmaceutical chemical plant. The entire process included development of a management of change policy, testing of change control procedures, training of personnel, followed by final implementation. The management of change process is initiated with a change permit, signed by those with designated responsibilities for making and reviewing the change. The process also includes provisions for a team evaluation of the change and for an independent review, which could include a process hazards review. As an aid to implementation a computerized management of change permit package was developed to reduce paperwork and processing time.

The specific process and equipment changes included in the program were:

- installation of new equipment, piping, or control systems;
- adding/deleting valves or piping;
- changes in pipe sizes;
- introduction of a new chemical or microorganism;
- changes in raw materials or waste streams;
- changes in hardware/software control systems;
- changes in the range of an instrument transmitter;
- new buildings or structures or modification thereof;
- generation of new waste streams;
- changes in storage tank contents or uses; and
- changes to equipment which could affect air contaminants or emissions.

E. IMPLICATIONS FOR OCCUPATIONAL SAFETY

Some kind of a management of change process should be part of any occupational safety management program. The guidance given by OSHA, EPA, and the Center for Chemical Process Safety, and directed to hazardous chemicals posing threats to workers and the population-at-large, applies to occupational safety management - though the formality and extent of application may be modified to suit the hazards.

For example, changes in mechanical production processes involving the addition of new machinery or equipment, although posing no toxic threat to workers or the public, should be carefully reviewed for occupational safety implications. The design of the equipment should be checked for worker protection concerns, the shut-offs should be located properly, the equipment should be installed safely, operators should be trained in safe operations, maintenance personnel should be trained to maintain the equipment, operating and maintenance procedures should be changed accordingly, plant drawings should be updated, etc. Prior to routine operation, the new equipment should be tested to ensure that everything is working as designed and that the revised procedures are correct. Any necessary preventive maintenance requirements should be added to the maintenance program. After operations have begun, safety should ensure that no unforeseen problems have developed.

MANAGING CHANGE: ASSESSMENT QUESTIONS

- Is there a written system to manage and assess changes in process:
  - chemicals?
  - technology?
  - equipment?
  - procedures?
  - facilities?
- Are proposed changes appropriately reviewed with regard to:
  - technical basis?
  - impact of change?
  - necessary time period for change?
- Are employees trained in the changes made prior to startup?
- Are all operating procedures and practices updated accordingly?
- Is there a way to detect and manage temporary changes including returning to original or

designed status at the end?
- Are forms or clearance sheets used to facilitate the processing of changes?
- Are changes in documents noted and made permanent?
- Are operating procedures or practices updated as needed?

XII. INCIDENT INVESTIGATIONS

A. OSHA REQUIREMENTS

The OSHA standard states that:

(1) The employer shall investigate each incident which resulted in, or could reasonably have resulted in a catastrophic release of highly hazardous chemical in the workplace.

(2) An incident investigation shall be initiated as promptly as possible, but not later than 48 hours following the incident.

(3) An incident investigation team shall be established and consist of at least one person knowledgeable in the process involved, including a contract employee if the incident involved work of the contractor, and other persons with appropriate knowledge and experience to thoroughly investigate and analyze the incident.

(4) A report shall be prepared at the conclusion of the investigation which includes at a minimum:

    (i)   Date of incident;

    (ii)   Date investigation began;

    (iii)   A description of the incident;

    (iv)   The factors that contributed to the incident; and,

    (v)   Any recommendations resulting from the investigation.

(5) The employer shall establish a system to promptly address and resolve the incident report findings and recommendations. Resolutions and corrective actions shall be documented.

(6) The report shall be reviewed with all affected personnel whose job tasks are relevant to the incident findings including *(applicable)* contract employees . . . .

(7) Incident investigation reports shall be retained for five years.

B. BACKGROUND

OSHA's background on investigation of incidents states that: Incident investigation is the process of identifying the underlying causes of incidents and implementing steps to prevent similar events from occurring. The intent of an incident investigation is for employers to learn from past experiences and thus avoid repeating past mistakes. The incidents for which OSHA expects employers to become aware and to investigate are the types of events which result in or could reasonably have resulted in a catastrophic release. Some of the events are sometimes referred to as "near-misses," meaning that a serious consequence did not occur, but could have.

Employers need to develop in-house capability to investigate incidents . . . . A team needs to be assembled . . . and trained in the techniques of investigation including how to conduct interviews of witnesses, needed documentation and report writing. A multi-disciplinary team is better able to gather the facts of the event and to analyze them and develop plausible scenarios as to what happened, and why. Team members should be selected on the basis of their training, knowledge and ability to contribute to a team effort to fully investigate the incident. Employees in the process area where the incident occurred should be consulted, interviewed or made a member of the team. Their knowledge of the events forms a significant set of facts about the incident which occurred. The report, its findings and recommendations are to be shared with those who can benefit from the information. The cooperation of employees is essential to an effective incident investigation. The focus of the investigation should be to obtain facts, and not to place blame. The team and the investigation process should clearly deal with all involved individuals in a fair, open and consistent manner.

OSHA cautions that the employer may reject proposals that are erroneous or infeasible, or may modify a recommendation that may not be as protective as possible or may be no more protective than a less complex or expensive measure. Resolution is when the employer adopted or justifiably declined a recommendation. When a recommendation is rejected, the employer must communicate this to the team and expeditiously resolve any subsequent recommendation.

INCIDENT INVESTIGATIONS: ASSESSMENT QUESTIONS

- For potentially catastrophic incidents, is there a ready, in-house incident investigation process with a knowledgeable, trained multi-disciplinary team?
- Are personnel trained to recognize incidents and near-incidents requiring investigation?
- Are incident investigations initiated as soon as possible (within 48 hours)?
- Do the investigations aim to focus on the facts and identify and correct the underlying causes of the incident?
- Are complete investigation reports prepared with findings and recommendations?
- Is there a system to address and resolve report findings and recommendations?
- Is the investigation report reviewed with all affected personnel when issued? periodically afterwards?
- Are incident investigation reports retained?

XIII. EMERGENCY PREPAREDNESS

A. OSHA REQUIREMENTS

The OSHA standard simply states that: The employer shall establish and implement an emergency action plan for the entire plant . . . . In addition, the emergency action plan shall include procedures for handling small releases. Employers . . . may also be subject to *(OSHA's)* hazardous waste and emergency response provisions . . . .

B. BACKGROUND
1. General Guidelines

OSHA's background on emergency preparedness states that: Each employer must address what actions employees are to take when there is an unwanted release of highly hazardous chemicals. Emergency preparedness or the employer's tertiary (third) lines of defense are those that will be relied on, along with the secondary lines of defense, when the primary lines of defense which are used to prevent an unwanted release fail to stop the release. Employers will need to decide if they want employees to handle and stop small or minor incidental releases. Or whether they wish to mobilize the available resources at the plant and have them brought to bear on a more significant release. Or whether employers want their employees to evacuate the danger area and promptly escape to a preplanned safe zone area, and allow the local community emergency response organizations to handle the release. Or whether the employer wants to use some combination of these actions. Employers will need to select how many different emergency preparedness or tertiary lines of defense they plan to have and then develop the necessary plans and procedures, and appropriately train employees in their emergency duties and responsibilities and then implement these lines of defense.

Employers at a minimum must have an emergency action plan which will facilitate the prompt evacuation of employees when there is an unwanted release of highly hazardous chemical. This means the employer will have a plan that will be activated by an alarm system to alert employees when to evacuate and, that employees who are physically impaired, will have the necessary support and assistance to get them to the safe zone as well. The intent of these requirements is to alert and move employees to a safe zone quickly. Delaying alarms or confusing alarms are to be avoided. The use of process control centers or similar process buildings in the process area as safe areas is discouraged. Recent catastrophes have shown that a large life loss has occurred in these structures because of where they have been sited and because they are not necessarily designed to withstand over pressures from shockwaves resulting from explosions in the process area.

2. Incidental Release Planning

Unwanted incidental releases of highly hazardous chemicals in the process area must be addressed by the employer as to what actions employees are to take. If the employer wants employees to evacuate the area, then the emergency action plan will be activated. For outdoor processes where wind direction is important for selecting the safe route to a refuge area, the employer should place a wind direction indicator such as a wind sock or pennant at the highest point that can be seen throughout the process area. Employees can move in the direction of cross wind to upwind to gain safe access to the refuge area by knowing the wind direction.

If the employer wants specific employees in the release area to control or stop the minor emergency or incidental release, these actions must be planned for in advance and procedures developed and implemented. Preplanning for handling incidental releases for minor emergencies in the process area needs to be done, appropriate equipment for the hazards must be provided, and training conducted for those employees who will perform the emergency work before they respond to handle an actual release. The employer's training program, including the Hazard Communication standard training, is to address the training needs for employees who are expected to handle incidental or minor releases.

3. Planning for Serious Releases

Preplanning for releases that are more serious than incidental releases is another important line of defense to be used by the employer. When a serious release of a highly hazardous chemical occurs, the employer through preplanning will have determined in advance what actions employees are to take. The evacuation of the immediate release area and other areas as necessary would be accomplished under the emergency action plan. If the employer wishes to use plant personnel such as a fire brigade, spill control team, a hazardous materials team, or use employees to render aid to those in the immediate release area and control or mitigate the incident, these actions are covered by 1910.120, the Hazardous Waste Operations and Emergency Response (HAZWOPER) standard. If outside assistance is necessary, such as through mutual aid agreements between employers and local government emergency response organizations, these emergency responders are also covered by HAZWOPER. The safety and health protections required for emergency responders are the responsibilities of their employers and of the on-scene incident commander.

Responders may be working under very hazardous conditions and therefore the objective is to have them competently led by an on-scene incident commander and the commander's staff, properly equipped to do their assigned work safely, and fully trained to carry out their duties safely before they respond to an emergency. Drills, training exercises, or simulations with the local community emergency response planners and responder organizations are means of obtaining better preparedness. This close cooperation and coordination between plant and local community emergency preparedness managers will also aid the employer in complying with the Environmental Protection Agency's Risk Management Plan criteria.

4. Emergency Control Centers

One effective way for medium to large facilities to enhance coordination and communication during emergencies for in-plant operations and with local community organizations is for employers to establish and equip an emergency control center. The emergency control center would be sited in a safe zone area so that it could be occupied throughout the duration of an emergency. The center would serve as the major communication link between the on-scene incident commander and plant or corporate management as well as with the local community officials. The communication equipment in the emergency control center should include a network to receive and transmit information by telephone, radio, or other means. It is important to have a backup communication network in case of power failure or one communication means fails. The center should also be equipped with the plant layout and community maps, utility drawings including fire water, emergency lighting, appropriate reference materials such as a government agency notification list, company personnel phone list, SARA Title III reports and material safety data sheets, emergency plans and procedures manual, a listing with the location of emergency response equipment, mutual aid information, and access to meteorological or weather condition data and any dispersion modeling data.

5. General OSHA Requirements

The emergency action plan must be in accordance with 29 CFR 1910.38(a) and include procedures for handling small releases. Certain provisions of the hazardous waste and emergency response standard 29 CFR 1910.120 (a), (p), and (q) may also apply. In complying with these provisions, it must be assured that:

The plan addresses:
* *Escape routes and procedures,*
* *Procedures for post-evacuation employee accounting,*
* *Duties and procedures of employees who:*
  Remain to operate critical equipment,
  Perform rescue and medical duties,
* *The names of persons or locations to contact for more action plan information.*

The plan must be written if the facility has more than 10 employees.

There must be a sufficient number of persons designated and trained to assist in the safe and orderly emergency evacuation of employees.

The plan must be reviewed with each employee covered by the plan: initially when the plan is developed; and whenever the plan changes; and responsibilities or designated action under the plan change; and whenever the plan is changed.

The emergency action plan must cover procedures for handling small releases.

6. OSHA Alarm Requirements

There should be an alarm system established which complies with 1910.165. In this case, the alarms must be:
* *Distinctive for each type of alarm.*
* *Capable of being perceived above the ambient noise and light levels by all employees in the affected portions of the workplace.*
* *Distinctive and recognizable as a signal to evacuate the work area or perform actions designated under the plan.*
* *Maintained in operating condition.*
* *Tested appropriately and restored to normal operating condition as soon as possible after test.*

Nonsupervised systems must be tested not less than every two weeks.

Supervised systems must be tested at least annually.
- *Supervised, maintained, and tested by appropriately trained persons.*
- *Unobstructed, conspicuous, and readily accessible, if they are manual alarm systems.*

7. Hazardous Chemical Requirements

In addition, the written emergency response plan must meet the requirements of 1910.120 if appropriate. 1910.120(q) address all facilities except TSD facilities and hazardous waste sites.

If employees are engaged in emergency response (except clean up operations) the plan must address the following:
- *Coordination with outside parties.*
- *Personnel roles, lines of authority, training, and communication.*
- *Emergency recognition and prevention.*
- *Safe distances and places of refuge.*
- *Site security and control.*
- *Evacuation routes and procedures.*
- *Decontamination.*
- *Emergency medical treatment and first aid.*
- *Emergency alerting and response procedures.*
- *Critique of response and follow-up.*
- *PPE and emergency equipment.*

People who are likely to discover hazardous releases must be trained to the first responder awareness level. People who will take defensive action in containing and controlling a release as part of the response must be trained to the first responder operations level. And, people who will take offensive action in containing and controlling the release as part of the response must be trained as hazardous materials (HAZMAT) technicians.

EMERGENCY PREPAREDNESS: ASSESSMENT QUESTIONS

- Is there an emergency action plan for the entire plant including procedures for small accidents and releases?
- Does the plan address:
  - escape procedures and routes?
  - post-evacuation employee accounting?
  - means of reporting emergencies?
  - individuals needed to remain or perform rescue and medical duties?
  - the employee alarm system?
- Are the scope and extent of employee involvement during emergencies specified?
- Does everyone know how and when to evacuate?
- Are there provisions for physically impaired personnel?
- Are all employees trained appropriately?
- Are response procedures verified to be implemented and adequate at least every three years?
- Are outside agencies prepared to respond?
- For hazardous chemicals, is there at least a plan for prompt evacuation activated by an alarm system?

XIV. COMPLIANCE AUDITS

The intent of this requirement is to require self-evaluation of the effectiveness of the PSM program by identifying deficiencies and assuring corrective actions.

A. OSHA REQUIREMENTS
(1) Employers shall certify that they have evaluated compliance with the provisions of this section at least every three years to verify that the procedures and practices developed . . . are adequate and are being followed.
If significant or numerous deficiencies are found, the frequency of audits may need to be increased.
(2) The compliance audit shall be conducted by at least one person knowledgeable in the process.
(3) A report of the findings of the audit shall be developed.
(4) The employer shall promptly determine and document an appropriate response to each of the findings of the compliance audit, and document that deficiencies have been corrected.
OSHA wants the appropriate response to be promptly documented and the corrective action documented as soon as possible after the action is taken.
(5) Employers shall retain the two (2) most recent compliance audit reports.

B. BACKGROUND
1. General Guidance
OSHA's background on compliance audits states that: Employers need to select a trained individual or assemble a trained team of people to audit the process safety management system and program. A small process or plant may need only one knowledgeable person to conduct an audit. The audit is to include an evaluation of the design and effectiveness of the process safety management system and a field inspection of the safety and health conditions and practices to verify that the employer's systems are effectively implemented. The audit should be conducted or lead by a person knowledgeable in audit techniques and who is impartial towards the facility or area being audited. The essential elements of an audit program include planning, staffing, conducting the audit, evaluation and corrective action, follow-up, and documentation.

2. Audit Planning
Planning in advance is essential to the success of the auditing process. Each employer needs to establish the format, staffing, scheduling, and verification methods prior to conducting the audit. The format should be designed to provide the lead auditor with a procedure or checklist which details the requirements of each section of the standard. The names of the audit team members should be listed as part of the format as well. The checklist, if properly designed, could serve as the verification sheet which provides the auditor with the necessary information to expedite the review and assure that no requirements of the standard are omitted. This verification sheet format could also identify those elements that will require evaluation or a response to correct deficiencies. This sheet could also be used for developing the follow-up and documentation requirements.

3. Audit Team
The selection of effective audit team members is critical to the success of the program. Team members should be chosen for their experience, knowledge, and training, and should be familiar with the processes and with auditing techniques, practices, and procedures. The size of the team will vary depending on the size and complexity of the process under consideration. For a large, complex, highly instrumented plant, it may be desirable to have team members with expertise in process engineering and design, process chemistry, instrumentation and computer controls, electrical

hazards and classifications, safety and health disciplines, maintenance, emergency preparedness, warehousing or shipping, and process safety auditing. The team may use part-time members to provide for the depth of expertise required as well as for what is actually done or followed, compared to what is written.

4. Audit Process

An effective audit includes a review of the relevant documentation and process safety information, inspection of the physical facilities, and interviews with all levels of plant personnel. Utilizing the audit procedure and checklist developed in the preplanning stage, the audit team can systematically analyze compliance with the provisions of the standard and any other corporate policies that are relevant. For example, the audit team will review all aspects of the training program as part of the overall audit. The team will review the written training program for adequacy of content, frequency of training, effectiveness of training in terms of its goals and objectives as well as to how it fits into meeting the standard's requirements, documentation, etc. Through interviews, the team can determine the employee's knowledge and awareness of the safety procedures, duties, rules, emergency response assignments, etc. During the inspection, the team can observe actual practices such as safety and health policies, procedures, and work authorization practices. This approach enables the team to identify deficiencies and determine where corrective actions or improvements are necessary.

An audit is a technique used to gather sufficient facts and information, including statistical information, to verify compliance with standards. Auditors should select as part of their preplanning a sample size sufficient to give a degree of confidence that the audit reflects the level of compliance with the standard. The audit team, through this systematic analysis, should document areas which require corrective action as well as those areas where the process safety management system is effective and working in an effective manner. This provides a record of the audit procedures and findings, and serves as a baseline of operation data for future audits. It will assist future auditors in determining changes or trends from previous audits.

5. Corrective Actions

Corrective action is one of the most important parts of the audit. It includes not only addressing the identified deficiencies, but also planning, follow-up, and documentation. The corrective action process normally begins with a management review of the audit findings. The purpose of this review is to determine what actions are appropriate, and to establish priorities, timetables, resource allocations and requirements and responsibilities. In some cases, corrective action may involve a simple change in procedure or minor maintenance effort to remedy the concern. Management of change procedures need to be used, as appropriate, even for what may seem to be a minor change. Many of the deficiencies can be acted on promptly, while some may require engineering studies or in-depth review of actual procedures and practices. There may be instances where no action is necessary and this is a valid response to an audit finding. All actions taken, including an explanation where no action is taken on a finding, need to be documented as to what was done and why.

It is important to assure that each deficiency identified is addressed, the corrective action to be taken noted, and the audit person or team responsible is properly documented by the employer. To control the corrective action process, the employer should consider the use of a tracking system. This tracking system might include periodic status reports shared with affected levels of management, specific reports such as completion of an engineering study, and a final implementation report to provide closure for audit findings that have been through management of change, if appropriate, and then shared with affected employees and management. This type of tracking system provides the employer with the status of the corrective action. It also provides the documentation required to verify that appropriate corrective actions were taken on deficiencies identified in the audit.

6. Audit Checklist

Dennison[8] presents a checklist of items to be covered in a typical PSM audit. The items are:

- written instructions for the operation;
- written instructions current;
- engineering drawings reflect modifications;
- process safety review of modifications;
- operators follow written instructions;
- equipment correct for the process;
- correct handling and storage of reactants;
- correct location of personal protective equipment;
- sufficient ventilation;
- process compatible electrical equipment;
- process hazardous materials identified;
- location of stored hazardous materials;
- quantities of stored hazardous materials;
- emergency procedures;
- emergency procedures and equipment training;
- identify potential errors in operation;
- identify potential consequences of errors in operation;
- operability of equipment safeguards;
- adequacy of explosion relief venting;
- proper direction of explosion reliefs;
- provisions for uncontrolled reactions;
- new process operator training;
- periodic training for existing operators;
- emergency training exercises;
- warranted chemical fire extinguishing systems;
- warranted explosion suppression system
- identify most likely possibilities of accidental releases;
- product contamination controls;
- storage of contamination sensitive bulk material;
- availability of critical spare parts;
- warranted emergency dump system; and
- critical maintenance procedures identified

COMPLIANCE AUDITS: ASSESSMENT QUESTIONS

- Have process safety management procedures and practices been periodically audited (at least every three years)?
- Do the audits include all PSM requirements?
- Do the audits include a field inspection and interviews with relevant staff?
- Is the lead auditor trained in audit techniques?
- Has the format of the audit been planned, including a checklist and sufficient sampling?
- Are audit team members chosen for their knowledge, experience, and familiarity with the process?
- Are audit results documented in a report and are the last two audit reports retained?
- Has a response to each finding been promptly determined and documented?
- Have audit deficiencies been followed up and verified as corrected?
- Is there a periodic review of audit results?
- Are audit results available to employees?

XV. SUMMARY - PROCESS SAFETY MANAGEMENT EVALUATION

Process Safety Management Topic	Rating
1. General Guidelines	
2. Employee Involvement	
3. Process Safety Information	
4. Process Hazards Analysis	
5. Operating Procedures and Practices	
6. Employee Training	
7. Contractors	
8. Pre-Startup Safety	
9. Mechanical Integrity	
10. Nonroutine Work Authorizations	
11. Managing Change	
12. Investigation of Incidents	
13. Emergency Preparedness	
14. Compliance Audits	
Total Safety Program Score	

A. PROCESS SAFETY MANAGEMENT RATING GUIDE

(Rate each topic 0 to 4. Consider the scope and breadth of deployment and the results achieved.)

4= Basically all of our people or activities meet the stated criteria or their intent with excellent results achieved.

3= Most of our people/activities meet the criteria or their intent with good results.

2= About half of our people/activities meet the criteria or their intent with positive results.

1= Some or a few of our people/activities meet the criteria or their intent with a few positive results.

0= None of our people/activities meet the criteria and no positive results are evident.

B. PROCESS SAFETY MANAGEMENT SCORING

Total Score   *Safety System Evaluation*

48-56   *PSM guidelines are completely built into our work safety system with very few or no gaps or problems.*

39-47	*Our company has very progressively built the PSM guidelines into safety with sound approaches and few gaps in deployment and integration but some fine tuning is possible in several areas.*
29-38	*Our company is progressive in building PSM guidelines into safety but gaps in deployment exist and refinements are still needed.*
19-28	*Our company has made progress in applying the PSM guidelines but significant gaps in deployment exist. Improvement is needed throughout.*
9-18	*Our company is just beginning to build the PSM guidelines into safety and requires substantial improvement in many ways.*
0-8	*Our company safety program is very traditional and reactive with limited success potential. The application of PSM guidelines to safety may be just beginning.*

XVI. REFERENCES

1. Process Safety Management of Highly Hazardous Chemicals - Compliance Guidelines and Enforcement Procedures, Compliance Directive CPC 2-2.45A CH-1, U. S. Occupational Safety and Health Administration, Washington, D. C., 1994.

2. Process Safety Management, OSHA-3132, U. S. Occupational Safety and Health Administration, Washington, D. C., 1994, Reprinted.

3. Process Safety Management Guidelines for Compliance, OSHA-3133, U. S. Occupational Safety and Health Administration, Washington, D. C., 1994, Reprinted.

4. **Stricoff, R. S.**, Safety Analysis and Process Safety Management, in *Risk Assessment and Management Handbook*, Kolluru, R. V., Ed., McGraw Hill, New York, NY, 1966, Ch. 8.

5. *Guidelines for Safe Process Operations and Maintenance*, The Center for Chemical Process Safety of the American Institute of Chemical Engineers, 1995, 226.

6. *Guidelines for Safe Process Operations and Maintenance*, 22-31.

7. **Jakubowski, J. A.**, Lessons learned: management of change, *Professional Safety*, 41.11, 29, 1996.

8. **Dennison, M. S.**, *OSHA and EPA Process Safety Management Requirements*, Van Nostrand Reinhold, New York, NY, 1994, 192-193.

Appendix A

BEHAVIOR-BASED SAFETY

I. DISCUSSION

Behavior-based safety is an approach which focuses on observable, measurable actions critical to safety. These critical work-related skills are discovered by applied behavioral analysis of data, including incident reports. Then processes are put in place which reinforce positive behaviors through a blame-free, no-fault, approach to behavioral and accident analysis, observation, and communication.[1]

The basis for safety improvement using behavior-based safety is that unsafe behaviors are the final common pathways of 80-95% of all accidents. They are the mass of preexisting behaviors that will certainly lead to accidents. Decreasing the number and frequency of unsafe behaviors will certainly lead to immediate decreases in incident and accident rates. Typical generic behaviors are such things as: body positioning/protecting, visual focusing, communicating, work pacing, moving objects, wearing protective equipment, complying with lockout/tagout, and complying with permits.

Typical measures of behavior safety success are: numbers of behavioral observations, percentage of employees volunteering to be observed, number of coaching sessions per week, and the percentage of safe behaviors per critical behavior category or work area.[2] Hidley and Krause[3] suggest that valid indications of safety performance are:

- the frequency of behavioral observations;
- the percentage of safe behaviors observed;
- involvement indicators and surveys; and
- safety-related maintenance information.

There is a direct relationship between behavior-based safety and TQM. In behavior-based safety:

- behaviors are observable and measurable, and as such are subject to the whole range of statistical analysis techniques;
- behavioral performance can be continuously improved;
- the focus is on observable behaviors, actions, and performance;
- management creates organizational consequences to reinforce critical behaviors;
- managers are held accountable for the quality of the safety culture;
- employees are empowered/involved in developing the lists of critical behaviors and in giving feedback; and
- behaviors are part of the facility's management system.

Although the practice of observation by employees is normal, there are other options available for the processes of conducting observations and giving feedback. Managers and supervisors can conduct observations and provide feedback, or employees can observe and report the results to supervisors - who provide feedback as part of their normal job duties. In all cases, the individuals need to be trained to provide appropriate and constructive feedback.[4]

In a behavior-based safety system there are:

- operational definitions of critical, site-specific, at-risk behaviors, where it is clear to the observer what the safe or at-risk behaviors are;

- ongoing, systematic observations of critical behaviors;
- regular, charted data and verbal feedback about observations;
- employee involvement in problem solving and action planning, on the basis of observed data; and
- subsequent observation and feedback on new measures.[5]

The steps which are usually followed in implementing behavior-based safety include:
- developing an inventory of operationally defined behaviors;
- fine tuning the inventory by staff review;
- using trained observers to establish a facility baseline;
- tracking and charting work group performance; and
- providing feedback.

Krouse[6] lists typical obstacles to high-quality observations as: resistance from employees, unions, and supervisors; over-familiarity with the work; unfamiliarity with the work or data sheet used to record observations; and not noticing the little but important things that could define safety.

Accident investigations are central to the behavior-based safety process. The goal is genuine accident prevention. The premise is that accidents typically result from known factors in the environment. Most traditional accident investigations err by looking for special, unique causes. Krouse and Russell[7] define accidents as "the unplanned result of a behavior that is likely part of a facility's culture." Accidents and incidents are treated the same, for as Krouse defines,[8] incidents are "the unplanned result of a behavior that (just) *happens not to cause* injury or damage." All accidents and injuries should be reported. In investigating accidents, management should use a systems approach, focusing on the root causes of loss. Accident investigations should ask: "Has the at-risk behavior been previously identified and operationally defined? Has the at-risk behavior been observed and at what frequency? Have employees been trained about this critical behavior? Is action plans in place to address these behaviors? If not, why not? Where is the accident prevention system failing?"

The A-B-C behavioral analysis technique is usually used to analyze problems and incidents. In this technique, the **A**ntecedents of **B**ehaviors lead to **C**onsequences. Antecedents (or activators) elicit behaviors because they signal or predict consequences. Consequences directly control behavior. Antecedents indirectly control behavior. McSween[9] states that "antecedents affect behavior because of the consequences." ABC analysis aims to discover the controlling consequences. In a complete ABC analysis for improvement, both the unsafe and safe behaviors should be analyzed. The old antecedents which triggered unsafe behaviors should be replaced by antecedents which promote safe behavior. For example, if signs were misplaced, they should be placed where they can be seen. If protective equipment was unavailable, it should be made easy to get. If workers didn't understand potential harm, they should be retrained. etc.

The magnitudes of accident consequences are determined by timing, consistency, and significance. Consequences that occur soon (S) are more controlling than those that occur late (L). Certain consequences (C) are more controlling than uncertain (U) ones. Positive (+) consequences are more controlling than negative (-) consequences. The stronger consequences are soon, certain, and positive.[10]

Examples of general antecedents are: signs, memos, instructions, policies, and mission statements. Work-related antecedents are: work demands, peer examples, and messages from management. Examples of general consequences are: praise, feedback, reprimand, and recognition. Work-related consequences are: comfort/convenience, work breaks, and peer/supervisor approval. Peer pressure is one of the most powerful consequences in an organization - it offers immediate, certain rewards to the worker who conforms.

Geller[11] points out that antecedents (activators) can be both internal and external. Internal activators can be goals that the individuals set for themselves. With internal activators there are corresponding internal consequences such as pride, dignity, feelings of accomplishment, increased self-esteem, etc. Internal activators for positive behavior are intrinsically better in that they can be relied upon even when the controlling external consequences are not available.

Corrective actions should be aimed toward altering the established consequences of behaviors. For example, if there is a history of incident/accident under reporting. Management should institute training, problem solving, ongoing behavioral observations, and no-fault accident investigations to alter the established consequences of accident/incident under reporting. Training would establish the need to report accidents and address peer criticism. Management and supervisor training would address fears of discipline and losing the safety prize. Safety meeting problem solving and action plans would address reporting and not being able to leave the job site. Positive new consequences should include - increased positive observations, positive feedback from observers, positive feedback from supervisors, improved safety ratings, and pride in performance.

Recognition for positive safety behaviors could be given for:
- submitting acceptable safety suggestions;
- heading safety meetings;
- participating in safety teams;
- completing all scheduled observations.[12]

Feedback to improve safety behaviors should be given in appropriate manners and at appropriate times. Geller[13] recommends that the best times for feedback are immediately after a behavior (positive or negative) or immediately preceding an anticipated behavior. The feedback should be nonjudgmental, should be focused and direct, and should be positive or negative as appropriate. However, care should be taken not to make the feedback too negative. As Geller[14] suggests, observations can be scheduled to allow time for employees to don personal protective equipment. This scheduling turns a potentially negative observation session into a positive one.

As a final note, it is only fair to observe that some of the premises of behavior-based safety are not accepted by all safety psychologists. Topf[15] for example, believes that focusing on employee attitudes is more important than just changing observed behaviors for driving continuous change. In his analysis, attitude-based changes are the ones which create self-initiating, self-perpetuating behavioral changes, and once enacted they do not depend on continuous monitoring to remain in force. Topf views attitude-based changes as inherently positive in nature, while the behavior-based changes are inherently negative. A worker with a good attitude behaves properly because he wants to remain healthy and whole. A worker responding only to behavior-based approaches might only be behaving properly to remain out of trouble.

II. TQM ASSESSMENT QUESTIONS MODIFIED TO INCLUDE BEHAVIOR-BASED SAFETY

In this appendix, the Total Quality Management questions presented in Chapter 2 have been modified to include the concepts and techniques of behavior-based safety. Modifications to the questions are shown in **bold, italic** print. Even where questions have not been changed or added, the intent of the questions should be interpreted from a behavior-based viewpoint.

A. PRODUCT AND CUSTOMER FOCUS
- Are the major focuses of all company activities, planning and culture, on the employee, on providing a safe, injury-free work environment and essential services, *on accident and injury prevention through behavior modification,* and on meeting employee expectations?

- *Are determining, analyzing, and improving upstream behavioral results prime focuses of company activities?*
- Do all the executives consider the employees to be customers?
- Have employee expectations and concerns regarding safety been determined in any formal and consistent manner?
- Do the executives know what the employees think about their safety programs?
- Do management and the workers have an expectation of zero injuries and continuous safety improvement?
- Does the safety policy refer to employee and management expectations?
- Does the organization communicate on safety with the external safety customers including the employee's family?

B. LEADERSHIP COMMITMENT:
- Are all our executives actively and fully committed and working toward a superior work safety culture?
- Do all our executives demonstrate their commitment and involvement on a routine basis?
- Do all executives really know about the company safety programs, safety record, and safety activities at this company?
- Are managers clearly responsible and accountable for safety performance?
- Do managers follow all the safety rules and requirements all the time?
- Do senior managers investigate accidents and conduct safety inspections, reviews, and audits?
- Do the executives provide the safety function with the resources needed (men, machines, methods, materials, media, motivation, and money) to promote and improve safety?
- Do all the managers go through the safety training? with their groups?
- Does the CEO or other top management get immediate notification of accidents?
- Do top management meetings include a review of safety performance?
- *Do the executives understand the principles and concepts of behavior-based safety?*
- *Have we established an organizational framework to reinforce critical behaviors?*

C. COMPANY CULTURE:
- Are company values, attitudes, and behaviors (management and employee) established to promote employee safety?
- Is safety a value at this company?
- Is there a strong safety culture established with no tolerance for unsafe practices?
- Do all of our executives and managers act as role models and promote this safety culture?
- Do all employees believe in the safety culture?
- Are safety procedures followed all the time?
- Is there a trusting relationship between management and labor?
- Do the employees believe that management responds ethically regarding safety issues?
- *Do the executives exhibit positive safety behaviors?*
- *Does employee peer pressure reinforce positive behaviors?*
- *Does the pervading culture shield the employees from blame and fear?*

D. EFFECTIVE COMMUNICATION:
- Do we use a variety of effective communications regarding safety including: routine training, meetings, memos, newsletters and postings, and informal discussions?
- In all of our communications is there reinforcement of the safety culture, vision, objectives, and commitment?

- Do all our employees freely discuss safety problems and offer safety suggestions to managers *and peers*?
- Does management positively and quickly respond to safety concerns and suggestions?
- Are employees protected from reprisals and harassment?
- Do managers and supervisors provide effective safety feedback to employees?
- *Do all employees communicate regarding safety in a positive, nonjudgmental way?*
- *Do employees provide specific, appropriate, and timely feedback on observations?*
- *Are the results of behavioral observations graphed and posted for the work area to see?*
- *Are these results discussed in safety meetings?*

E. ORGANIZATIONAL AND EMPLOYEE KNOWLEDGE:
- Do our organization and all of our workforce exhibit the kinds of knowledge and characteristics necessary to achieve work safety objectives?
- Do we generate and use new ideas for safety management?
- Do we adapt ideas and techniques from others to enhance safety?
- Do all our people understand the quality and work safety programs and their core principles and do they have all the skills necessary to work safely?
- Do our people strive to learn more and develop their skills?
- Do employees participate in the training program other than as students?
- Is there a minimum safety skills and knowledge level set for employees?
- *Do our managers, supervisors, and employees understand the principles of behavior-based safety and know how to apply them in the work environment?*
- *Does our organization really understand the statistical nature of accident data?*
- *Can the employees observe and interpret safety behaviors?*
- *Can our employees participate in accident/incident analyses with a behavioral viewpoint?*

F. EMPLOYEE EMPOWERMENT:
- Are all our employees prepared for and empowered to fully participate in work planning and safety, including making in-field safety decisions?
- Do all our employees have the authority, given by management, to act independently to meet safety expectations?
- Are all employees given expectations, guidelines, resources, and skills necessary to act independently?
- Do employees inspect for hazards? analyze jobs? develop safe work practices? train new employees in safety? and assist in accident investigations?
- Do employees have stop work authority if unsafe conditions develop?
- *Do the employees fully participate in determining the critical safety behaviors, observing and providing feedback on these behaviors, and determining corrective actions?*

G. EMPLOYEE RESPONSIBILITY AND EXCELLENCE:
- Are all employees prepared for and sharing in responsibility for a safe work environment?
- Do all the employees acknowledge work safety and all its objectives, agree with work safety policies and programs, accept their roles in the organization, and act responsibly as safe employees?
- Do I see all the employees working safely all the time?
- Do employees participate in audits? inspections? incident analyses?
- Do employees analyze the safety performance of co-workers and coach them for improvement?

- Do employees learn and use quality tools for safety?
- Are employees expected to contribute safety improvement ideas?
- *Do all employees participate in the behavior-based safety programs by observing behaviors and positively responding to safety comments?*

H. MANAGEMENT BY FACT:

- Is our work safety management system based on specific, timely, achievable, and realistic performance measures, on positive performance indicators, and on a framework of information, measurement, data, and analysis?
- Do all of our executives review safety results?
- Are measures other than accident statistics used as a primary performance measure?
- Are accident statistics used to assess performance in a statistically valid way?
- Does the workforce understand what statistical analyses are valid?
- Do our performance measures include employee satisfaction and morale? safety-related behavioral competencies? safety skill and knowledge acquisition? safety suggestions implemented?
- Do we measure the real costs of accidents and injuries?
- Are safety achievements measured and displayed to the workforce?
- *Has a list of critical safety behaviors been developed and are they systematically and routinely observed, measured, tracked, and reported?*
- *Are other behavioral measures used to track performance?*

I. LONG-RANGE VIEWPOINT:

- Are there demonstrated strong future orientation and long-range planning for safety and employee development with a willingness to make long-term work safety commitments?
- Do we have a strategic plan for safety development?
- Do we plan and commit for future employee development and for improvements in workplace safety?
- Do we keep aware of regulatory changes and initiatives and do we plan for them?
- Does our budget reflect the continuing commitment of money and resources for training and safety?
- *Does the company have a long-range view of continuous behavioral improvement?*
- Does the company avoid short-term reactive approaches to accidents, incidents, and safety problems in favor of long-term corrective actions?

J. STATISTICAL PROCESS CONTROL:

- Are statistical process analyses and methods used to manage and control workplace safety in all departments and processes?
- Are accident and injury statistics analyzed in a statistically valid way?
- Are any of the following charts or diagrams used for statistical measurement and control: pareto charts, run charts, scatter diagrams, histograms, control charts, tally charts?
- Do the managers and employees know what these charts/diagrams are and how to use them?
- Does everyone have a good understanding of accident statistics and what are valid results?
- *Are behavior-based safety observation results analyzed statistically?*
- *Are behavior-based safety observations reported to the employees statistically?*

K. STRUCTURAL PROBLEM SOLVING:

- Are effective analysis and problem solving techniques used to identify problems, and analyze accidents and safety problems?

- What specific problem analysis and solution techniques have personnel been trained to use?
- Are injury and illness trends analyzed, so that patterns with common causes can be identified and prevented?
- Are there documented examples showing how the problem solving techniques have been used?
- Have the employees been trained in the basic steps of problem solving?
- Are any of the listed voting techniques used to decide among alternative courses of action?
- *Is the Antecedent-Behavior-Consequence analysis technique used to understand and develop solutions for behavior-based safety problems?*
- *Does incident/accident analysis seek to determine the common, behavioral causes of accidents?*

L. THE BEST TECHNIQUES:
- Has a work safety baseline been determined to identify the hazards and control methods used?
- Are world-class (best, better than required, etc.) work safety techniques from other organizations identified and applied to improve safety?
- What efforts are made to identify best/better techniques from both inside and outside the organization for potential improvements?
- Is a complete benchmarking process used to conduct the studies and implement actions?
- Have improvements been made in this way?
- Do all employees strive to help identify areas for improvement?

M. CONTINUOUS IMPROVEMENT:
- Is there a record of continuous safety improvement through the cycle of problem identification and analysis, development and implementation of corrective recommendations, review of results and development of effective controls?
- What kinds of safety improvements have been made?
- Are the results of improvement efforts tracked and reviewed?
- How often are changes made in the basic work processes?
- *Has the record of unsafe behaviors shown continuous improvement?*
- How often are changes made in the basic work processes?

N. QUALITY MANAGEMENT:
- Has a comprehensive quality and safety management system including policies, processes, procedures, and standards been developed and installed?
- Are all the facets of the management system effectively implemented?
- How well can the managers and employees describe the work safety management system and how it is organized?
- *Do our standards and goals address safety behavior improvement?*

O. QUALITY PLANNING:
- Is there an effective planning program to identify the critical and common cause factors of accidents and illnesses and prevent accidents prior to occurrence for all of our work processes and activities?
- Are any specific analysis tools/techniques used in planning for work safety?
- Is staff trained to effectively use the quality planning tools?
- Is the responsibility for planning spread throughout the organization?
- *What critical safety factors have been identified for the major job/process categories?*

P. ASSESSMENT AND PLANNING:
- Have our development and implementation of the work safety management system included a complete program assessment, strategic planning to develop new courses of action by executives, and tactical planning with safety projects and objectives?
- Have we developed safety vision, mission, and values statements, and planned to implement a continuous improvement and effective process management structure?
- Have our vision, mission, and values statements been clearly communicated to staff?
- Have we planned to establish safety projects, committees, and teams to implement the safety vision and mission?
- Was the complete organization represented in the safety assessment and planning?
- *Have the employees been involved in safety assessment and planning, particularly the lists of critical safety factors and the assessment implementation?*

Q. IMPLEMENTATION AND ORGANIZATION:
- Have we established an effective safety organization and work safety management system?
- Do we have safety projects, project teams, and new processes, systems, and techniques?
- Are all departments aligned with internal requirements, department mission statements, measurements of satisfaction, and agreements on meeting expectations?
- Is there a strong and effective corporate safety council, chaired by a senior executive?
- Are work safety teams organized and used for involvement, agility, and an ownership focus?
- *Do all departments measure and have goals for improving safety behaviors?*

R. CULTURAL CHANGE:
- Has there been a transformation of the work safety culture to one in which all basic TQM principles are embodied so that quality and safety are equal in importance to product delivery and cost?
- Is there is a management-led improvement process and do all employees participate with a focus on common goals *including exhibiting positive safety behaviors?*
- Is safety as important to the executives as are production, quality, and profit?
- Is safety as important to the employees as production and getting the work done on time?
- Has management instituted safety excellence awards?
- Has our work culture changed to one of safety by prevention rather than inspection?

S. RECOGNITION AND REWARD:
- Is our recognition and rewards process established for employee safety and for improvements in the safety of the workplace and processes?
- Is everybody in the organization (managers, supervisors, and employees) clearly held accountable for meeting their safety responsibilities?
- To what extent do performance appraisals focus on job performance *and adherence to safety behaviors*?
- *Are positive safety behaviors routinely recognized by peers and supervisors?*
- *Are supervisors evaluated on how well the observation process is conducted?*
- Do our compensation systems have any linkage with work safety performance?
- Is my input sought on performance appraisals?
- Are our safety rewards team-based rather than individual-based?
- Are safety rewards equivalent to our quality and production rewards?

T. LEADERSHIP DEVELOPMENT:
- Have we developed and educated all of our employees for safety leadership to help implement the work safety management system?

- Do we develop leaders who: share visions, motivate employees, and ensure that individuals are assessed based on performance?
- Do we develop leaders who: focus on results, build employee commitment, effectively lead meetings, and create positive working relationships?
- Is safety leadership only left to the safety group?
- Do employees: identify concerns? report concerns to others? suggest solutions? consult with others? and follow up on corrective actions?
- Do supervisors: continually demonstrate safety behavior? show a commitment to safety? and reward and praise safety excellence?
- Do our leaders: have a safety vision? have confidence in the staff? take risks to improve safety? make hard safety decisions? develop other safety leaders? influence others to work safely? and continually communicate the safety message?
- *Do the employees exhibit leadership in observing and counseling on unsafe behaviors?*

U. TEAM BUILDING:
- Have we established safety (and project) teams?
- Are all team members educated in effective team work concepts, including the importance of two-way communications?
- Have we taught the team members the skills needed to successfully function on a team?
- Have we carefully selected strong team leaders?
- Are teams and team members clearly accountable for their performance on the team and as team members?
- Do the teams successfully work on safety projects?
- Does each safety team have input or a member from the safety function?
- *Have we used teams to develop the behavior-based implementation process?*
- *Have we used teams to analyze work functions and develop lists of critical safety behaviors?*
- *Have we used teams to analyze and correct safety problems?*

V. HIRING AND PROMOTING:
- Do we make a major effort to only hire those who will fit in with the organization's quality/safety culture?
- Are our hiring interviews tough and demanding?
- Are the line managers personally and strongly involved in hiring?
- Does the orientation program stress the safety culture, safety programs, and safety expectations?
- Are all of our promotion decisions based on acceptance of our safety culture and safety performance, equally with other criteria?
- Are all promotions in alignment with safety results, leadership through safety and quality, *exhibition and promotion of positive safety behaviors*, human resources management, teamwork, and corporate safety values?
- Does the safety function have any input on promotion and hiring decisions?
- *Are our hiring decisions based on the expectation of positive safety behaviors?*

W. MANAGEMENT READINESS:
- Are all our managers (especially first-line supervisors) trained in the skills essential for the safety culture? These skills are: observing and interpreting safety behaviors, counseling (or coaching) for safety, running results-oriented safety meetings, and conducting safety-oriented performance appraisals.

- Are all our supervisors ready to perform the functions of: analyzing the work they supervise to identify unrecognized potential hazards, maintaining physical protection in their work areas, reinforcing employee training on the nature of potential hazards and on needed protective measures, **and reinforcing correct work behaviors?**
- Are the supervisors ready and able to coach for improved safety through communicating, observing, analyzing, changing, and helping?
- Do the managers and supervisors effectively apply these skills?

X. TRAINING:

- Does our training program provide everyone with the bases and elements for all work safety characteristics and concepts, and for the continual implementation and improvement in work safety programs?
- Does our safety training program include the following elements: work safety concepts, safety/quality tools and techniques, and leadership?
- Does training include both technical and personal skills?
- Are managers, supervisors, and employees used as trainers?
- Is training targeted to the job and applied to the job?
- *Have all personnel been trained to understand and use behavior-based safety ideas and techniques?*
- *Have managers, supervisors, and observers been trained to effectively observe and coach on safety behaviors?*

III. REFERENCES

1. **Krause, T. R., Hidley, J. H., and Hodson, S. J.**, *The Behavior-Based Safety Process*, Van Nostrand Reinhold, New York, NY, 1990.
2. **Geller, E. S.**, Barriers to breakthrough performance, *Industrial Safety and Hygiene News*, June, 1996, Internet: www.safetyonline.net/ishn/9606/behav.htm.
3. **Krause**, *The Behavior-Based Safety Process*, 47-48.
4. Improve your safety program with a behavioral approach, *The Quality Safety Edge*, Internet: www.safetyonline.net/hse/qse/process.htm.
5. **Hidley, J. H. and Krause, T. R.**, Behavior-based safety: paradigm shift beyond the failures of attitude-based programs, *Professional Safety*, 39.10, 28, 1994.
6. **Krause, T. R.**, *Employee-Driven Systems for Safe Behavior*, Van Nostrand Reinhold, New York, NY, 1995. 125-126.
7. **Krause, T. R. and Russell, L. R.**, The behavior-based safety approach to proactive accident investigation, *Professional Safety*, 39.3, 22, 1994.
8. **Krause**, *Employee-Driven Systems for Safe Behavior*, 169.
9. **McSween, T. E.**, *The Values-Based Safety Process*, Van Nostrand Reinhold, New York, NY, 1995, 238.
10. **Krause**, *The Behavior-Based Safety Process*, 21.
11. **Geller, E. S.**, How to motivate behavior for lasting results, *Industrial Safety and Hygiene News*, March, 1996, Internet: www.safetyonline.net/ishn/9603/behav.htm.
12. **McSween**, *The Values-Based Safety Process*, 167.
13. **Geller, E. S.**, 20 guidelines for giving feedback, *Industrial Safety and Hygiene News*, July, 1996, Internet: www.safetyonline.net/ishn/9607/behav.htm.
14. **Geller, E. S.**, Two common coaching mistakes, *Industrial Safety and Hygiene News*, May, 1996, Internet: www.safetyonline.net/ishn/9605/behav.htm.
15. **Topf, M. P. and Petrino, R. A.**, Change in attitude fosters responsibility for safety, *Professional Safety*, 40.12, 24, 1995.

Appendix B

AUDITING

The material in this appendix is intended to provide background material for performing audits.

Audits are generally conducted to the ISO auditing[1] or similar guidelines.[2] The ISO guidelines are common denominators for auditing in business organizations of all kinds. The methodology in these guidelines is universal, and is discussed first. Then an enhanced process, an "Advanced Audit" process for preparing for audits is described. This process results in the development of logically sound audit plans and well-supported audit results. The appendix also includes a discussion of interviewing, material on developing findings, conclusions, and recommendations, and conducting exit interviews.

I. AUDIT DEFINITION AND PURPOSE

ISO-10011-1 defines a quality audit as follows: "An audit is a systematic and independent examination to determine whether quality activities and related results comply with planned arrangements and whether these arrangements are implemented effectively and are suitable to achieve objectives." Audits can be applied to systems, processes, products, programs, or services. They can be internal, performed by the company, or they can be external, performed by suppliers or hired consultants. Their purposes can also be to provide an opportunity for improvement, and can be performed to meet regulatory requirements.

When performed for internal systems review, as an integral part of management review activities, audits are generally scheduled based on the importance and risk presented by the activity. Audits are not strictly black-and-white processes. In order to determine implementation effectiveness, the auditor must interpret data obtained from a variety of sources to make an informed judgment about the quality system. In addition, to draw conclusions and make effective recommendations, the auditor must be able to analyze findings in an organizational context and select the most appropriate and saleable corrective suggestions.

II. AUDIT CONDUCT

Audits generally follow a standard format. First the general purpose and schedule of the audit are determined. Then the audit leader and team members are selected. Specific audit planning follows. Actually conducting the audit usually involves an opening meeting, document reviews, observations and interviews, a period of analysis, a closing meeting, and then submittal of an audit report. The audit also should include follow-up to ensure that any corrective actions are implemented.

As described in ISO-10011-1, audits are typically conducted by an audit team headed by a lead auditor. The auditors' responsibilities are to: comply with audit requirements, communicate and clarify audit requirements, plan and carry out their assigned responsibilities, document audit observations, report audit results, verify corrective action effectiveness, retain and safeguard audit documents, and cooperate with and support the lead auditor.

The lead auditor is responsible for the audit. He usually has management experience and has the authority to make final audit decisions. The lead auditor also assists in the selection of the audit team members, prepares the audit plan, represents the audit team with management, and submits the audit report.

The lead auditor defines audit assignments, establishes auditor qualifications, plans the audit, prepares working documents, immediately reports critical nonconformances, reports major audit problems, and reports audit results clearly, conclusively and without undue delay.

ISO-10011-1 specifies that the auditors: remain within the audit scope, exercise objectivity, collect and analyze evidence relevant and sufficient to permit the drawing of conclusions regarding the quality system, remain alert to any indications of evidence that can influence audit results and require more extensive auditing, be able to judge if the quality system is implemented effectively and if the quality system documentation is adequate, and act ethically at all times.

A. THE AUDIT SCOPE

If conducted for management or system review purposes, the audit scope is usually specified by management or by the quality department. Decisions are made regarding which quality system elements, physical locations, and organizational activities are to be audited with the assistance of the lead auditor. The scope and depth of the audit should meet specific information needs. The standards or documents with which the auditee's quality system is required to comply should be specified by the client. Sufficient objective evidence should be available to demonstrate the operation and effectiveness of the auditee's quality system. The resources committed to the audit scope should be sufficient to meet the intended scope and depth.

B. AUDIT PLANNING

Planning for the audit involves selecting a skilled and qualified team; confirming the audit's objective and scope and the specific quality requirements; identifying information sources; planning the audit program; confirming the program; and developing checklists.

ISO-10011-1 recommends that the audit plan should be flexible in order to respond to observations or problems that may develop. The specific plan should include: audit objectives and scope, identification of individuals to be interviewed, identification of reference documents, identification of audit team members, the expected time for each major audit activity, the schedule of audit meetings, any confidentiality requirements, audit report distribution and the expected date of issue. Audit plans should be reviewed to determine if they are logically planned, if they have achievable objectives, and if they are properly scheduled.

The information sources should be chosen to ensure a balanced view of the company's operation. Sources of information from which to develop the audit checklists and program include: quality manual and procedures, management priorities, quality problems, previous audits - outstanding corrective actions, product information, and the experience of the auditors.

The documents required for the audit and to document and report results may include check-sheets, forms to report observations, and forms to document supporting evidence for conclusions. Working documents should not restrict additional audit activities or investigations which may become necessary. Any confidential or proprietary information in the working documents should be safeguarded.

C. AUDIT EXECUTION
1. The Opening Meeting

The purpose of the opening meeting is to establish two-way communication and start the audit off in the right way. The meeting should start on time and last no longer than thirty minutes.[3] During this meeting, the audit team is introduced to senior management, the audit scope and objectives are reviewed, audit methods and procedures are summarized, communication links are established, needed resources and facilities are confirmed to be available, the time and date of any interim and closing meetings are confirmed, and any unclear audit details are clarified.

2. Collecting Evidence

During the audit, evidence is collected through interviews, document examinations, and activity and condition observations. Any information that suggests the presence of nonconformances should be noted and investigated. Interview information should be corroborated. This can be done by getting the same information from independent sources, such a physical observations, measurements, or records.

3. Observations

Audit observations and findings should be documented and verified. During and after the audit, all observations and findings should be reviewed to determine which are to be reported as nonconformances. These should be clearly and concisely documented, and should be supported by evidence. All nonconformances should be referenced to specific requirements of the standards or criteria used. During the audit, observations should be reviewed with the responsible manager and all nonconformances should be verified by management.

When recording a nonconformance, enough details need to be provided so that the auditee can reexamine the observations later. Nonconformance details should include: where it was found, an exact observation of the facts surrounding the discrepancy, the reason why it is a nonconformance, sufficient traceability references. In writing up a nonconformance, it is good to: use local terminology, make the information easily retrievable for future reference, make it helpful to the auditee, and make it concise - yet complete.

4. Team Meetings

Daily team meetings should be held at the end of each day's auditing. The team members review their activities, observations, tentative conclusions, and problems. The audit schedule or plan may be changed as a result. The purpose of the daily meeting is to bring everyone up-to-date, to convey audit information and to make on course corrections as necessary.[4]

5. Closing Meeting

The closing or exit meeting represents the opportunity for the audit team to verbally present their findings, conclusions, and recommendations, and to get feedback as to their validity or usefulness. The audit team meets with senior management and the functional managers just after the audit and prior to writing the audit report. The meeting should begin with a summary of the audit. During the meeting, the lead auditor presents the team's observations, findings, conclusions, and recommendations. He notes the significance of the findings and conclusions, and discusses the appropriateness of any recommendations. Audit team members may participate as necessary or appropriate to describe their specific tasks or observations or to clarify any points of contention. Any future actions on the part of the audit team should be described. The audit team should keep a record of the closing meeting.

6. Audit Report

The audit report is the written record of the audit, its results, including recommendations, and any necessary follow-up activities. It reflects the tone and content of the audit and presents a balanced picture by identifying positive and negative aspects of the company's systems. ISO-10011-1 specifies that the audit report should describe: the audit scope and objectives, details of the audit plan, the audit team members and any company representatives, the dates of the audit, the specific organization audited, the reference documents used as auditing standards, noncompliance observations, positive and negative findings and conclusions related to the quality system and its objectives, and any recommendations for corrective action deemed appropriate. The audit report should be dated and signed by the lead auditor, and should have a predetermined distribution list.

7. Corrective Action Follow-Up

Corrective action closes the loop in the audit process. Once any nonconformances are reported or once recommendations are offered, the auditee should determine and initiate corrective action. There should be an agreed-to time period for corrective action and a method to verify that the action was completed. Follow-up audits may be performed to verify that the nonconformances were corrected. A corrective action follow-up procedure should include the following: identification and agreement as to the details of the nonconformance; agreement on the corrective action; agreement on the resolution timetables and dates; corrective action implementation; effectiveness evaluation, and a follow-up audit to confirm completion.

D. AUDIT PROBLEMS

Parnswith[5] lists typical audit problems to avoid. These are:
- inadequate planning and preparation;
- inadequate communication with the auditee prior to the audit;
- lack of clearly defined scope;
- lack of understanding of the standards against which the audit is being performed;
- lack of properly trained auditors including technical and interviewing skills;
- too prescriptive in evaluating corrective action responses; and
- failure to reevaluate implementation of corrective actions.

III. AUDITOR QUALIFICATION AND CHARACTERISTICS

Auditors should be adequately qualified to perform the audit.[6] Qualification can include having a specified professional and quality background or certification, a knowledge of the function to be audited, a knowledge of the standards to be used/applied, and a knowledge of management systems, and the possession of a high degree of ethics and integrity.

Technically, the auditor should know the technical standards applicable to the product, service, or function of the organization. These can be codes or process specifications. The auditor should know cost accounting, statistical techniques (sampling, interpreting, confidence levels), and should know diagnostic techniques (problem solving, troubleshooting).

The auditor should have requisite personality traits such as leadership, confidence, composure, and independence. He should also be flexible and thorough, formal and documented. The auditor should understand investigative techniques, should be able to communicate well orally and in writing, should be able to constructively critique, and should be decisive about audit results. He should also be able to present and discuss recommendations, and alternatives and their impact on the audited organization.

IV. THE ADVANCED AUDIT PROCESS

The advanced audit process is a generalized process for logically constructing audit or assessment plans. This process has been used to create targeted assessment plans to investigate many types of management and programmatic problems. The audit plans permit **Findings**, **Conclusions**, and **Recommendations**, to be developed where the **Conclusions** have sufficient justification to buttress the corrective **Recommendations**.

The basic process in developing a logically constructed audit plan is straightforward although the individual steps could be complex and time consuming. The plan could begin with an **Area of Focus** or a **Major Issue** or even tentative recommendations based on prior knowledge. Plan development could be top-to-bottom or bottom-to-top with ideas generated individually or by group brainstorming and organized in affinity or interrelationship diagrams.

A. STRUCTURE OF ADVANCED AUDITS

In outline form, the advanced audit plan consists of:

- An **Area of Focus**
- Several **Major Issues** (see the definition of terms)
- **Alternatives** (at least one for each Major Issue)
- **Hypotheses** (several for each Major Issue)
- **Assumptions** (several for each Hypothesis)
- **Key Questions** (one for each Assumption)
- A **Data Package**

The results generally consist of:

- **Findings**
- **Conclusions**
- **Recommendations**

The structure of an advanced audit plan is shown in Figures 1 and 2.

These terms are defined below.

- **Area of Focus**: Typically this is a broad area of concern or interest. Such an area may be training or maintenance.
- **Major Issue**: These are issues or questions which are necessary to assess the adequacy of the Area of Focus. Major Issues are themselves broad and one or two may comprise the total agenda of the typical inspection.
- **Hypotheses** (Working Hypotheses): These are tentative statements developed for a Major Issue. The Hypotheses should be sufficient to adequately address the Issue and answer the Major Issue questions.
- **Assumptions**: These are propositions or statements that something is true. Generally a number of Assumptions are developed for each Hypothesis and if verified provide it full support.
- **Key Questions**: These are the Assumptions posed in general question form to promote development of a detailed audit plan. The answer to a Key Question would generally be a Finding.
- **Data Package**: The sum total of interview requirements, documents to be reviewed, data needed, auditor tasks including analyses, exhibits, and observations, interview guides with general and probe questions needed to address the Key Questions.
- **Alternatives**: These are propositions or situations offering a choice. In planning an audit, Alternatives may be developed based on prior knowledge as tentative or possible recommendations. Alternatives may be developed after the audit based on audit results, offering a choice of possible solutions to a problem.
- **Findings**: Findings are facts related to the Major Issue, which surface during an evaluation of a specific area of focus. Verified Assumptions are Findings.
- **Conclusions**: These are the auditor's opinions of conditions which are developed after interpreting findings, and are suggestive of actions. Verified Hypotheses are Conclusions.
- **Recommendations**: These are the auditor's suggestions or action statements for addressing problems described by the Conclusions.

Note that care needs to be taken in phrasing all inspection plan statements so that the scope is clearly defined and the supportive statements provide necessary and sufficient support.

B. AUDIT PLAN DEVELOPMENT

Generally the audit begins with an Area of Focus or a Major Issue. Once these have been clearly drawn, development of Hypotheses and Assumptions proceeds in an iterative fashion which may be:

196

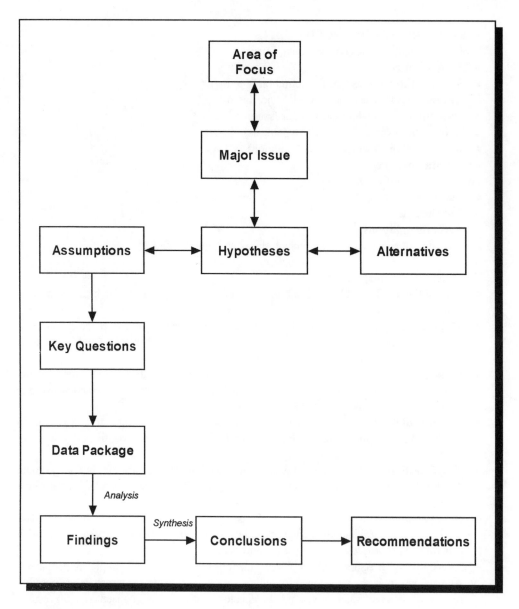

Figure 1. The Structure of Advanced Audits. This figure shows how all the aspects of the Advanced Audit are related.

- **Top-Down**; wherein supporting Assumptions for each Hypothesis are developed by analysis of what is needed to provide fully supportive statements.
- **Functionally**; wherein the Assumptions are derived by analysis of the functions required to achieve accomplishment of the Hypotheses. For example, if the Hypothesis concerns the adequacy of spare parts, the Assumptions might concern purchasing, storage, receipt, inspection, etc.
- **Bottom-Up**; wherein proposed statements of fact, Assumptions, are developed and then grouped into tentative Hypotheses which address a Major Issue.

197

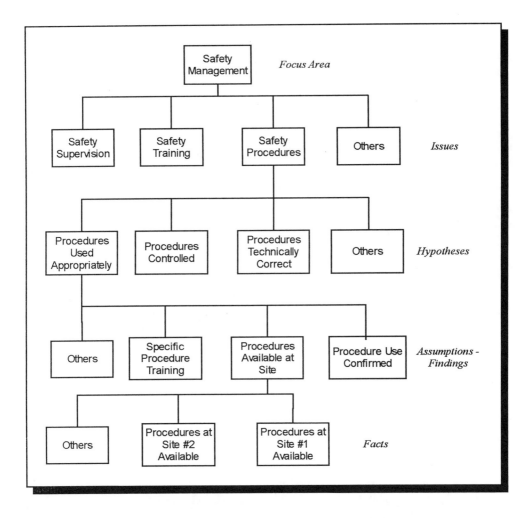

Figure 2. Example of an Advanced Audit where the Safety Management Focus Area works down to the level of where copies of procedures are located in the field. In this case, procedures are at sites 1 and 2, thus the Finding is that procedures are available at operating sites in the field.

- **Brainstorming**; wherein a group lists important functional and management aspects of an Area or Issue and then these are grouped into logical and supportive Hypotheses and Assumptions.

While there is no one right way to develop an audit plan, what is important is that:
- The plan elements are fully supportive of higher elements
- The plan elements are specific and concrete rather than vague and amorphous in character
- The plan elements include critical management or performance principles
- The plan is complete
- The bottom-line effectiveness is assessed

There are no "magic numbers" of issues or components that are needed for an effective audit plan. The number of issues that can be reviewed is limited by the duration of the audit and the sufficiency of the audit plan. Plans should not be disjointed or overblown which waste data collection and analysis time and resources. The development must proceed in a questioning self-critical manner which focuses on the real issues and what is really needed to prove a point or justify a conclusion.

The planning format sets the stage for collecting the factual information required to draw solid conclusions and make meaningful recommendations. It helps maximize the collection of relevant information and minimize the collection of irrelevant information. Using the logic diagram, work is planned and controlled and gaps in information are highlighted. A sample audit plan for Operational Worker Safety is shown in Figure 3.

Sample Audit Plan

Area of Audit: Operational worker safety

Major Issues: Have operator training programs and practices been designed and implemented to fully support operational worker safety.

Working Hypotheses: 1. Control, monitoring, and design of on-the-job (OJT) efforts have minimized operator injuries.
2. Additional hypotheses.

Assumptions: 1. There is a formal OJT program.
2. The OJT program is consistently applied.
3. The OJT program contains appropriate schedules and milestones.
4. There are a minimum number of operator injuries.
5. Etc.

Key Questions: 1. Are OJT control programs formalized and do they effectively track OJT?
2. Etc.

Data Package: Copy of OJT control system.
Exhibit and analysis: The % of OJT accomplished by each supervisor.
Inspector tasks: Interview list, analysis requirements, observations to make.
Source of inputs: Field, office, etc.

Interview Guide: Probing questions. Example: Has the night shift staff had OJT on a specific system?

Figure 3. A sample audit plan for operational worker safety.

During the audit, assumptions will become findings, hypotheses will become conclusions, and alternatives will become recommendations. Depending on the audit, initial assumptions are modified, refined, and strengthened into findings. As findings are developed, the initial hypotheses are either supported or reshaped into firm conclusions. The conclusions in turn, lead to recommendations. The results of applying this plan are shown in Figure 4.

Sample Audit Results

Area of Audit: Operational worker safety

Major Issues: Have operator training programs and practices been designed and implemented to fully support operational worker safety.

Recommendations: Develop methods to give increased control of OJT to assure consistent application.

Conclusions: 1. **While formal programs are available, control of OJT has not been sufficient to assure consistent application. Missed OJT may have contributed to operator injuries**.
      2. Additional conclusions.

Findings: 1. There is a formal OJT program.
      2. The OJT program is applied **inconsistently.**
      3. The OJT program contains appropriate schedules and milestones.
      4. There are an **above average** number of operator injuries.
      5. Etc.

Key Questions: 1. Are OJT control programs formalized and do they effectively track OJT?
      2. Etc.

Data Package: Copy of OJT control system.
Exhibit and analysis: The % of OJT accomplished by each supervisor.
Inspector tasks: Interview list, analysis requirements, observations to make.
Source of inputs: Field, office, etc.

Interview Guide: Probing questions.
Example: Has the night shift staff had OJT on a specific system?

Figure 4. An example of a completed audit plan where the negative findings are shown in bold.

V. INTERVIEWING

The interview process is one in which planned, sequenced, open-ended, and probing questions are asked, answers are listened to attentively and responded to appropriately, and notes are taken to document the information. One vital characteristic of a good interview is that the interviewer is listening 80% of the time while talking only 20%. The purpose of an interview is to gather information. It must be planned for and organized in a manner that allows the free, accurate, and ordered flow of information and data with an objective in mind. A fact gathering interview consists of five major segments: the preparation phase; the start of the interview; the body of the interview; the close of the interview; and the evaluation of the interview.

A. THE PREPARATION PHASE

A properly prepared interview enhances the process by indicating to the interviewee that you are aware of his organization and structure, its basic operating procedures, and the current operating mode. It will also aid in directing the interview in a manner which helps keep things on track. In preparing, keep in mind the purpose of the interview and the results expected. Prepare an interview guide, starting with broad questions, phrasing questions as you will ask them, organizing the questions logically, determining who to interview, and screening out unnecessary questions.

Note that open-ended questions tend to minimize defensiveness on the part of the interviewee and will often generate a broader response than close-ended questions. Since there is no implied need to agree or disagree, open-ended questions avoid telegraphing the expected answer. On the other hand, close-ended questions require a short, specific response. They are useful when specific information such as names and dates is needed, or when following up on vague answers. In addition, clarifying questions are used to ensure an understanding of what the interviewee is saying. Clarifying techniques include: probing or using follow-up questions; paraphrasing or repeating and rewording important points; and summarizing or recapping and repeating a set of major points.

B. STARTING THE INTERVIEW

There are points of professionalism that encourage a positive tone throughout the interview process. These are: promptness; following internal client protocols; informing superiors of interviews with subordinates; introducing yourself and the reason for the interview; noting that the purpose is to audit the system not the client; starting with broad questions; establishing a level of rapport; communicating at the same responsibility level and knowledge; striving for confidence, frankness, informality, pleasantness, and interest; and using appropriate body language to show that you are listening

C. THE INTERVIEW

The interview proper should flow from the start in a smooth evolutionary manner. Interviewees should feel comfortable when answering questions. This will promote the free flow of information and perhaps open new issues. Interviewers should ask open-ended questions when possible. They should write down examples when necessary, listen to what the interviewee is saying and is not saying, respond to answers as appropriate, deviate from and revise questions as necessary, and write legibly, completely, but concisely. They should make sure key details are accurately captured, quoting directly where the flavor is important or meaning is critical. Overall, they should maintain control of the interview.

D. THE CLOSE OF THE INTERVIEW

As the interview starts to close, the interviewer should complete the process in a succinct and direct manner. The interviewer should make it obvious that the interview is over, thank the interviewee for helping, and should leave the door open to return.

E. THE INTERVIEW EVALUATION

At the end of the interview, the information gathered must be evaluated for usefulness. The interviewer should consider if he obtained everything that was planned. He should fill in notes immediately after the interview, separating facts from impressions. Any areas or inconsistencies that require additional information or investigation should be noted. Impressions of the interviewee should also be documented. Finally, the interviewer should identify key issues and findings, determining the meaning of the data in terms of the audit objectives.

F. INTERVIEWING FAULTS

The following faults can hinder an interview and reduce its overall effectiveness:[7,8]

- Not talking as a professional.
- Having a poor introduction.
- Using biased or leading questions.
- Insufficient note taking.
- Losing control.
- Not listening enough.
- Asking too many closed-ended questions.
- Insufficient probing.
- Not summarizing enough to ensure accuracy.

G. THE INTERVIEW GUIDE

The interview guide prepares the interviewer by aiding in: developing interview questions based on key questions, developing the opening statement, testing the questions to ensure the objective of the inspection is reached, and revising the inspection scope as situations and conditions change.

In preparing the interview guide: develop interview questions from the key questions for the project, use broad open-ended questions where appropriate to allow the interviewee room to expand, develop questions to get at key questions which cannot be asked because of sensitivity (do not be unnecessarily blunt), develop questions to gain access to printed or other factual data material, use probing questions for details and clarification, ask for examples or explanations when necessary, and construct the questions carefully. A sample interview guide is shown in Figure 5.

Prepare a broad introductory question. This establishes a frame of reference and sets the tone the interview will follow. It will also help put the interviewee at ease. Prepare a brief opening statement. This allows you to describe your purpose for being there and the topics you will cover. Briefly describe yourself. This helps put you on equal terms with the interviewee. Test the interview guide. Eliminate or revise unproductive questions. Improve the wording of questions as needed. Clarify subquestions and probes.

There is an art to selecting the most proper way of phrasing and sequencing questions. For example, in evaluating how personnel performance is handled, you might start with an open-ended question like - Describe how you measure performance. In a more direct way you could use a sequence such as: Who are you best performers? How do you know? Show me.

VI. DEVELOPING FINDINGS AND CONCLUSIONS

A. THE FACT SHEET

The fact sheet is a tool for collecting data and facts during the interviews and inspection, and for reviewing these facts to arrive at a finding. A sample fact sheet is shown in Figure 6.

B. DEVELOPING FINDINGS

Review the completed fact sheet to identify major ideas and construct an answer to the key question. This is similar to constructing a topic sentence for a paragraph. The answer summarizes the central theme of the supporting facts. The answer may not include all the supportive ideas and facts. The answer should not include anything which is not listed on the completed fact sheet.

Draft the answer (the finding) in a single sentence and list this at the bottom of the fact sheet. A finding is stated as a complete sentence. A finding is a descriptive answer to the key question. A finding is not a judgmental answer which evaluates the merits of the situation or provides a "yes" or "no" answer. Revise and sharpen the finding sentence to get down to the exact meaning, to replace abstract words with concrete words, and to eliminate unnecessary words.

1. Describe your present business.
    What services do you render?
    Who are your major customers?
    How do you sell?
    What has been the history of growth?
    Profits?
    Who is your major competition?
    How large are your competitors?
    What has been their growth?
    How profitable have they been?
    How are they organized?
    What major plans are available?

2. What is your organization?
    Whom do you report to?
    Who reports to you?
    What changes, if any will be made
    when the subsidiary companies are
    systematically merged?

3. What are your responsibilities?
    Do you have a written position
    description?
    Approximately what percent of the time
    do you spend on each function?

4. How is your personal performance
    measured?
    What measures are used?

    What records are available which cover
    these measures?

5. What are your working relationships
    with others in the organization?
    What services do you supply? To
    whom?
    What services do you require? From
    whom?

6. Who are your strongest backup men?
    What are their names, positions, ages,
    and salary histories?
    Can you give me a brief appraisal of
    each?
    Do you have position descriptions for
    these subordinate positions?

7. How do you measure the performance
    of your immediate subordinates?
    What measures do you use?
    What records show these results?

8. What suggestions do you have for
    improving the organization?
    What are your major problem areas?
    Are there functions which are now
    needlessly duplicated elsewhere in the
    organization?

Figure 5. A sample interview guide

C. DEVELOPING CONCLUSIONS

Analyze the findings and draw conclusions about their meaning in relation to the nature and scope of the study. First consider, "Taken together, what do these findings tell me about the problems or issues of the study?" Next, group related findings and ask, "What do these related findings tell me?" Last, compare findings against each other and ask, "What do these diverse findings, when compared to each other, tell me?"

VII. DEVELOPING RECOMMENDATIONS

A well-developed audit plan followed by effective implementation during the audit becomes meaningless if the recommendations which result are inappropriate or not accepted by the client. Both the style of the auditor and the form and soundness of the recommendation are important considerations in developing recommendations. In general, there are two types of styles appropriate to most advanced audits. These styles are the instructor and the catalyst. The instructor provides expert information and recommends a course of action. This requires a high degree of accurate diagnosis. The catalyst acts as a change agent. The catalyst helps the group to solve its own problems, makes the group aware of the consequences, and passes on appropriate alternatives. The catalyst style is most often used to create an interest in change. However, the two styles can be used alternatively in an audit.

Recommendations are the auditor's tools for solving the problems described by the conclusions. Recommendations can be presented in several different forms: providing a specific solution, outlining alternative solutions, or defining the need for further study necessary in order to determine the proper course of action.

Prior to issuing any final recommendations, the auditor must answer several questions. Have all the alternatives been analyzed? Will the client accept them as presented? How does the auditor's point of view differ from the clients? Dealing with these questions will often promote the auditor's reevaluation of wording or approach to one which would be more acceptable.

As a final step, each recommendation should be tested for logic, feasibility, and creativity. Does the individual recommendation adequately describe action and appropriately address benefits? Do all the recommendations, when taken together, adequately cover the major issue addressed in the audit? Are the recommendations sound? Do they resolve the problem and are reasonably logical when inserted into the discussion of the audit's conclusions? Have all hipshooting statements been eliminated? Is each recommended course of action feasible given the client's technical, economic, and regulatory constraints? Does it fit with the client's corporate structure?

The following sequence of steps can be followed in developing alternatives and recommendations.

A. ESTABLISH OBJECTIVES FOR IMPROVEMENT

Given the nature of the assignment, what objectives do we need to accomplish? Objectives grow out of the analysis of the problem. Findings give positive and negative dimensions of the problem solution. Conclusions provide direction and impetus to seek improvement. Objectives for improvement are concerned with such topics as profit, sales, personnel, production, etc. Describe the "solution in principal" - your idea of the ideal solution.

B. DEVELOP ALTERNATIVE APPROACHES TO OBTAINING THE OBJECTIVES

The objective is to develop a number of approaches which can move us to our goal. Build from a positive foundation by asking, "What findings do I have that I want to retain and/or improve on?" Remove negative aspects of the situation by asking, "What negative findings can be removed while meeting my objectives?" Consider approaches which have been used on similar assignments. This requires disciplined creativity to develop alternative approaches within the nature and scope of the job.

C. SELECT THE BEST ALTERNATIVE

The objective is to select the most workable, timely, practical, and acceptable solution in terms of the client's needs and resources. No single alternative will be a perfect fit; each has hidden negative factors and risks. There is a need to construct a set of evaluation criteria which considers: the client's attitudes and abilities, the costs involved, the personality factors, the disruption involved, the timing involved, and the effects on profits, legal implications, and morale. Select the alternative that promises the most advantages with the fewest disadvantages.

D. DETERMINE THE INSTALLATION STEPS

Once the best alternative is selected, it must be installed. First, identify the key installation steps. Then determine the timing and sequence of the installation steps. Then, assign responsibility and designate performance standards for the installation. Finally, review the results and verify that the objectives were met.

> **Task:** Copy directly from the work plan.
>
>
> **Key Question:** Copy directly from the planning work sheet.
>
>
> **Relevant Facts:**
>
> List bits and pieces of your answer to the question.
>
> > Able to reduce your handling of working papers.
> >
> > Lets you put your interview notes away.
> >
> > Major bits and pieces can be used to construct your answer on the bottom of the sheet - your finding.
> >
> > Remaining facts are used to support your major heading.
>
> Good communications tool - copies can be distributed to team members.
>
> Fact sheet contains "hard facts" - what can be said with certainty about this question.
>
> Able to see when you have enough facts to answer the question.
>
> Able to manage another person's data collection.
>
> Able to spot holes and conflicts in the data.
>
>
>
> **Finding:** A simple, complete, nonjudgmental sentence which helps define the present situation and/or tests the validity of the hypothesis.

Figure 6. The Fact Sheet - How It Is Used.

E. SELL THE SOLUTION

Develop a set of positive recommendations. Relate the recommendations to the objectives which the client is seeking. State the benefits to be obtained. List interim steps if the client cannot immediately achieve objectives.

VIII. EXIT INTERVIEWS

Successfully communicating findings to auditees means: relating new information to that already known by the audience; providing a logical order so the audience can follow the thoughts from a point to point; reinforcing the message and audience response by using appropriate repetition and summarization.

The primary expectations in communicating with clients are to:

- Discuss the problems and possible solutions.
- Provide a mechanism to achieve corrective action.
- Agree on proper corrective action.
- Develop a good working relationship.
- Provide a channel for exchanging pertinent information.

The key element to successful communication is the credibility of the speaker. If someone is able to successfully challenge a 'fact', the basis for the rest of the presentation becomes suspect.

Conclusions may be positive. The area or program audited may satisfy all requirements. No corrective action is required. On the other hand, the audit conclusion may be that problems exist which call for action in order to bring a program into compliance or to correct the underlying causes of the problem. The conclusion should be worded in such a manner that suggests possible corrective actions. However, all corrective actions need to be tailored to fit the special situation at the facility/organization being audited. Generic solutions are not always available.

In preparing for a presentation, questions or objections to the findings or conclusions should be anticipated. These should be worked into the presentation before they can be voiced. This will help keep control of the presentation and defuse potentially volatile situations.

A. EXIT INTERVIEWS

The exit interview or meeting is an important part of the advanced audit process. These interviews allow for open discussion of findings, conclusions, and resulting recommendations. Several benefits are derived from the exit interview process: The client is notified of the perceived problem soon after it has been identified. The opportunity exists to begin taking corrective action, even before more senior management has received the final report or letter. The auditor has another opportunity to verify the accuracy and validity of the findings and conclusions. Where conflict exits, the auditor is in a position to begin negotiations with the client to reach a reasonable, mutually agreed upon solution.

B. OPEN DISCUSSION OF ITEMS

Evading the key issues and points of conflict will not get suggestions or recommendations implemented. The client will not know the specific concerns unless they are presented openly. Soft discussions and roundabout logic waste time and irritate client staff. Such efforts may also build barriers between the auditors and the clients.

Clients may not like what they hear. However, an honest, accurate, straightforward presentation of findings, conclusions, and recommendations helps develop and maintain a good working relationship with the client. The descriptive mode of results presentation should be used rather than the critical, evaluative mode.

C. LISTENING

The exit interview should not be a monologue. A good dialogue needs to be actively pursued with the client. The client should understand and accept all concerns raised, and the auditor should also understand and appreciate the client's concerns and position. This is accomplished by listening carefully and maintaining an open mind. The active listening skills described previously should be used.

The tone of voice used by the client in responding to the audit presentation is important. If there is irritation, conflict may follow. The auditor should be aware of resistance to change. The auditor should also try to stay on the subject. If people try to sidetrack the discussion, the auditor should redirect the discussion back to the topic at hand.

REFERENCES

1. Guideline for auditing quality systems - Part 1: Auditing, ISO-10011-1, 1st Edition, International Organization for Standardization, Geneva, Switzerland, 1990.
2. **Middleton, D.**, Internal Quality Audits, in *The ISO-9000 Handbook*, 2nd Edition, Peach, R. W., Edition, Irwin Professional Publishing, Fairfax, VA, 1995, 150-163.
3. **Parnswith, B. S.**, *Fundamentals of Quality Auditing*, ASQC Quality Press, Milwaukee, WI, 1995, 21.
4. **Parnswith**, *Fundamentals of Quality Auditing*, 28.
5. **Parnswith**, *Fundamentals of Quality Auditing*, 33.
6. **Mills, C. A.**, *The Quality Audit: A Management Evaluation Tool*, ASQC Quality Press, Milwaukee, WI, 1989, Ch. 7.
7. **Flagg, J. C., Kerr, D. S., and Smith, L. M.**, Conducting effective interviews, *Internal Auditing*, 10.3, 41, 1995.
8. **Pratt, R. C.**, Interview or Inquisition: Successful Communication Techniques, in *The ISO-9000 Handbook*, 2nd Edition, Peach, R. W., Ed., Irwin Professional Publishing, Fairfax, VA., 1995, 164-168.

INDEX

DATE DUE

DEMCO 38-297